MODULAR
FUNCTION SPACES

MONOGRAPHS AND TEXTBOOKS IN
PURE AND APPLIED MATHEMATICS

30. *J. S. Golan*, Localization of Noncommutative Rings (1975)
31. *G. Klambauer*, Mathematical Analysis (1975)
32. *M. K. Agoston*, Algebraic Topology: A First Course (1976)
33. *K. R. Goodearl*, Ring Theory: Nonsingular Rings and Modules (1976)
34. *L. E. Mansfield*, Linear Algebra with Geometric Applications: Selected Topics (1976)
35. *N. J. Pullman*, Matrix Theory and Its Applications (1976)
36. *B. R. McDonald*, Geometric Algebra Over Local Rings (1976)
37. *C. W. Groetsch*, Generalized Inverses of Linear Operators: Representation and Approximation (1977)
38. *J. E. Kuczkowski and J. L. Gersting*, Abstract Algebra: A First Look (1977)
39. *C. O. Christenson and W. L. Voxman*, Aspects of Topology (1977)
40. *M. Nagata*, Field Theory (1977)
41. *R. L. Long*, Algebraic Number Theory (1977)
42. *W. F. Pfeffer*, Integrals and Measures (1977)
43. *R. L. Wheeden and A. Zygmund*, Measure and Integral: An Introduction to Real Analysis (1977)
44. *J. H. Curtiss*, Introduction to Functions of a Complex Variable (1978)
45. *K. Hrbacek and T. Jech*, Introduction to Set Theory (1978)
46. *W. S. Massey*, Homology and Cohomology Theory (1978)
47. *M. Marcus*, Introduction to Modern Algebra (1978)
48. *E. C. Young*, Vector and Tensor Analysis (1978)
49. *S. B. Nadler, Jr.*, Hyperspaces of Sets (1978)
50. *S. K. Segal*, Topics in Group Rings (1978)
51. *A. C. M. van Rooij*, Non-Archimedean Functional Analysis (1978)
54. *L. Corwin and R. Szczarba*, Calculus in Vector Spaces (1979)
53. *C. Sadosky*, Interpolation of Operators and Singular Integrals: An Introduction to Harmonic Analysis (1979)
54. *J. Cronin*, Differential Equations: Introduction and Quantitative Theory (1980)
55. *C. W. Groetsch*, Elements of Applicable Functional Analysis (1980)
56. *I. Vaisman*, Foundations of Three-Dimensional Euclidean Geometry (1980)
57. *H. I. Freedman*, Deterministic Mathematical Models in Population Ecology (1980)
58. *S. B. Chae*, Lebesgue Integration (1980)
59. *C. S. Rees, S. M. Shah, and C. V. Stanojević,* Theory and Applications of Fourier Analysis (1981)
60. *L. Nachbin*, Introduction to Functional Analysis: Banach Spaces and Differential Calculus (R. M. Aron, translator) (1981)
61. *G. Orzech and M. Orzech*, Plane Algebraic Curves: An Introduction Via Valuations (1981)
62. *R. Johnsonbaugh and W. E. Pfaffenberger*, Foundations of Mathematical Analysis (1981)
63. *W. L. Voxman and R. H. Goetschel*, Advanced Calculus: An Introduction to Modern Analysis (1981)
64. *L. J. Corwin and R. H. Szcarba*, Multivariable Calculus (1982)
65. *V. I. Istrătescu*, Introduction to Linear Operator Theory (1981)
66. *R. D. Järvinen*, Finite and Infinite Dimensional Linear Spaces: A Comparative Study in Algebraic and Analytic Settings (1981)

67. *J. K. Beem and P. E. Ehrlich,* Global Lorentzian Geometry (1981)
68. *D. L. Armacost,* The Structure of Locally Compact Abelian Groups (1981)
69. *J. W. Brewer and M. K. Smith, eds.,* Emmy Noether: A Tribute to Her Life and Work (1981)
70. *K. H. Kim,* Boolean Matrix Theory and Applications (1982)
71. *T. W. Wieting,* The Mathematical Theory of Chromatic Plane Ornaments (1982)
72. *D. B. Gauld,* Differential Topology: An Introduction (1982)
73. *R. L. Faber,* Foundations of Euclidean and Non-Euclidean Geometry (1983)
74. *M. Carmeli,* Statistical Theory and Random Matrices (1983)
75. *J. H. Carruth, J. A. Hildebrant, and R. J. Koch,* The Theory of Topological Semigroups (1983)
76. *R. L. Faber,* Differential Geometry and Relativity Theory: An Introduction (1983)
77. *S. Barnett,* Polynomials and Linear Control Systems (1983)
78. *G. Karpilovsky,* Commutative Group Algebras (1983)
79. *F. Van Oystaeyen and A. Verschoren,* Relative Invariants of Rings: The Commutative Theory (1983)
80. *I. Vaisman,* A First Course in Differential Geometry (1984)
81. *G. W. Swan,* Applications of Optimal Control Theory in Biomedicine (1984)
82. *T. Petrie and J. D. Randall,* Transformation Groups on Manifolds (1984)
83. *K. Goebel and S. Reich,* Uniform Convexity, Hyperbolic Geometry, and Nonexpansive Mappings (1984)
84. *T. Albu and C. Năstăsescu,* Relative Finiteness in Module Theory (1984)
85. *K. Hrbacek and T. Jech,* Introduction to Set Theory, Second Edition, Revised and Expanded (1984)
86. *F. Van Oystaeyen and A. Verschoren,* Relative Invariants of Rings: The Noncommutative Theory (1984)
87. *B. R. McDonald,* Linear Algebra Over Commutative Rings (1984)
88. *M. Namba,* Geometry of Projective Algebraic Curves (1984)
89. *G. F. Webb,* Theory of Nonlinear Age-Dependent Population Dynamics (1985)
90. *M. R. Bremner, R. V. Moody, and J. Patera,* Tables of Dominant Weight Multiplicities for Representations of Simple Lie Algebras (1985)
91. *A. E. Fekete,* Real Linear Algebra (1985)
92. *S. B. Chae,* Holomorphy and Calculus in Normed Spaces (1985)
93. *A. J. Jerri,* Introduction to Integral Equations with Applications (1985)
94. *G. Karpilovsky,* Projective Representations of Finite Groups (1985)
95. *L. Narici and E. Beckenstein,* Topological Vector Spaces (1985)
96. *J. Weeks,* The Shape of Space: How to Visualize Surfaces and Three-Dimensional Manifolds (1985)
97. *P. R. Gribik and K. O. Kortanek,* Extremal Methods of Operations Research (1985)
98. *J.-A. Chao and W. A. Woyczynski, eds.,* Probability Theory and Harmonic Analysis (1986)
99. *G. D. Crown, M. H. Fenrick, and R. J. Valenza,* Abstract Algebra (1986)
100. *J. H. Carruth, J. A. Hildebrant, and R. J. Koch,* The Theory of Topological Semigroups, Volume 2 (1986)

101. *R. S. Doran and V. A. Belfi,* Characterizations of C*-Algebras: The Gelfand-Naimark Theorems (1986)
102. *M. W. Jeter,* Mathematical Programming: An Introduction to Optimization (1986)
103. *M. Altman,* A Unified Theory of Nonlinear Operator and Evolution Equations with Applications: A New Approach to Nonlinear Partial Differential Equations (1986)
104. *A. Verschoren,* Relative Invariants of Sheaves (1987)
105. *R. A. Usmani,* Applied Linear Algebra (1987)
106. *P. Blass and J. Lang,* Zariski Surfaces and Differential Equations in Characteristic p > 0 (1987)
107. *J. A. Reneke, R. E. Fennell, and R. B. Minton.* Structured Hereditary Systems (1987)
108. *H. Busemann and B. B. Phadke,* Spaces with Distinguished Geodesics (1987)
109. *R. Harte,* Invertibility and Singularity for Bounded Linear Operators (1988).
110. *G. S. Ladde, V. Lakshmikantham, and B. G. Zhang,* Oscillation Theory of Differential Equations with Deviating Arguments (1987)
111. *L. Dudkin, I. Rabinovich, and I. Vakhutinsky,* Iterative Aggregation Theory: Mathematical Methods of Coordinating Detailed and Aggregate Problems in Large Control Systems (1987)
112. *T. Okubo,* Differential Geometry (1987)
113. *D. L. Stancl and M. L. Stancl,* Real Analysis with Point-Set Topology (1987)
114. *T. C. Gard,* Introduction to Stochastic Differential Equations (1988)
115. *S. S. Abhyankar,* Enumerative Combinatorics of Young Tableaux (1988)
116. *H. Strade and R. Farnsteiner,* Modular Lie Algebras and Their Representations (1988)
117. *J. A. Huckaba,* Commutative Rings with Zero Divisors (1988)
118. *W. D. Wallis,* Combinatorial Designs (1988)
119. *W. Więsław,* Topological Fields (1988)
120. *G. Karpilovsky,* Field Theory: Classical Foundations and Multiplicative Groups (1988)
121. *S. Caenepeel and F. Van Oystaeyen,* Brauer Groups and the Cohomology of Graded Rings (1988)
122. *W. Kozlowski,* Modular Function Spaces (1988)

Other Volumes in Preparation

MODULAR
FUNCTION SPACES

Wojciech M. Kozlowski

University of Southern California
Los Angeles, California

MARCEL DEKKER, INC. New York and Basel

Library of Congress Cataloging-in-Publication Data

Kozlowski, Wojciech M.
    Modular function spaces.

    (Monographs and textbooks in pure and applied
mathematics ; v. 122)
    Bibliography: p.
    Includes index.
    1. Function spaces. I. Title. II. Series.
QA326.K69 1988   515.7'3   88-20269
ISBN 0-8247-8001-9

MARCEL DEKKER, INC.
270 Madison Avenue, New York, New York 10016

Current printing (last digit):
10 9 8 7 6 5 4 3 2 1

PRINTED IN THE UNITED STATES OF AMERICA

To   Julian   Musielak

Preface

During their everyday work many mathematicians have to deal with the problem of equipping a given vector space with a suitable topology. Since the solution is not always obvious, several techniques have been developed. One of them, the theory of modular spaces, has proved to be especially efficient whenever function spaces are involved. Roughly speaking, modulars are the functionals that generalize norms and F-norms. In addition to the general theory of modular spaces, some more specialized cases have been thoroughly investigated, Orlicz spaces among them. We may observe, however, many situations where the theory of modular spaces is too general, while the methods of Orlicz spaces are too restrictive. To fill this gap is the aim of the theory of modular function spaces and of this book in particular.

The author's hope is that the theory and applications presented here will provide useful techniques that can be directly applied or at least can serve as an inspiration for developing new notions and methods.

This work was prepared while the author was a Fulbright Scholar visiting the California Institute of Technology in Pasadena, California. The author would like to express his sincere gratitude to the Fulbright Foundation for supporting the project and to the Department of Mathematics at Caltech for providing excellent research facilities. The author would like to acknowledge comments, suggestions, and critical remarks from many friends.

Wojciech M. Kozlowski

Contents

MODULAR
FUNCTION SPACES

1. Introduction

First attempts to generalize the classical function spaces of the Lebesgue type L^p were made in the early 1930's by W. Orlicz and Z. Birnbaum in connection with orthogonal expansions. Their approach consisted in considering spaces of functions with some growth properties different from the power type growth control provided by the L^p- norms. Namely, they considered the function spaces defined as follows:

$$L^\phi = \left\{ f : \mathbb{R} \rightarrow \mathbb{R} \; ; \; \exists \; \lambda > 0 : \int_\mathbb{R} \phi\Big(\lambda|f(x)|\Big) \, dx < \infty \right\},$$

where $\phi : [0,\infty] \rightarrow [0,\infty]$ was assumed to be a convex function increasing to infinity, i.e. the function which to some extent behaves similarly to power functions $\phi(u) = u^p$ (for precise definitions see Chapter 4). Later on, the assumption of convexity for Orlicz functions ϕ was frequently omitted. Let us mention two typical examples of such functions:

$$\phi(u) = e^u - 1 \qquad \text{and} \qquad \phi(u) = \ln(1 + u).$$

The possibility of introducing the structure of a linear metric

1

space in L^ϕ as well as the interesting properties of these spaces and many applications to differential and integral equations with kernels of nonpower types were among the reasons for the development of the theory of Orlicz spaces, their applications and generalizations for more than half of the century. We would like to stress that recently a new interest in classical Orlicz spaces is emerging in connection with problems of convexity, the Boyd indices and rearrangement invariant function spaces (see e.g. J. Lindenstrauss and L. Tzafriri, [2]).

We may observe two principal directions of further development. The first one is a theory of Banach function spaces initiated in 1955 by W.A.J Luxemburg and then developed in a series of joint papers with A.C. Zaanen. The main idea of that theory consists in considering a function space L of all measurable functions $f : X \rightarrow \mathbb{R}$ such that $\|f\| < \infty$, where (X,Σ,μ) is a measure space, $M(X,S)$ denotes the space of all strongly measurable functions acting from X into a Banach space S and $\|\cdot\|$ is a function norm, i.e. a functional $\|\cdot\| : M(X,\mathbb{R}) \rightarrow [0,\infty]$ such that,

  (i) $\|f\| = 0$ if and only if $f = 0$ μ-a.e.;

  (ii) $\|\lambda f\| = |\lambda| \|f\|$ for scalars λ;

  (ii) $\|f+g\| \leq \|f\| + \|g\|$;

  (iv) $\|f\| \leq \|g\|$ whenever $|f(x)| \leq |g(x)|$ μ-a.e.

Luxemburg and Zaanen considered conditions for completeness of such normed spaces. They also developed a theory of duality of Banach function spaces as well as many interesting applications to integral equations and other chapters of analysis.

The other way, also inspired by the successful theory of Orlicz spaces, is based on replacing the particular, integral form of the nonlinear functional, which controls the growth of members of the space, by an abstractly given functional with some good properties. This idea was the basis of the theory of modular spaces initiated by H. Nakano in 1950 in connection with the theory of order spaces and redefined and generalized by W. Orlicz and J. Musielak in 1959. Let us give a brief account of some basic facts of their theory.

Let $\mathfrak{X}$ be a vector space over $\mathbb{K}$ ($\mathbb{K} = \mathbb{C}$ or $\mathbb{K} = \mathbb{R}$). A functional $\rho: \mathfrak{X} \to [0,\infty]$ is called a pseudomodular, if for arbitrary f and g, elements of $\mathfrak{X}$, there holds :

(1) $\rho(0) = 0$;

(2) $\rho(\alpha f) = \rho(f)$ for every $\alpha \in \mathbb{K}$, $|\alpha| = 1$,

(3) $\rho(\alpha f + \beta g) \leq \rho(f) + \rho(g)$ for $\alpha, \beta \geq 0$, $\alpha + \beta = 1$.

If we replace (3) by

$(3')$ $\rho(\alpha f + \beta g) \leq \alpha^s \rho(f) + \beta^s \rho(g)$

        for $\alpha, \beta \geq 0$, $\alpha^s + \beta^s = 1$, $s \in (0,1]$, then the pseudomodular ρ is called s-convex (convex for $s = 1$). If in place of (1) there holds

(1′) $\rho(0) = 0$ and $\rho(\lambda f) = 0$ for all $\lambda > 0$ implies $f = 0$,

then ρ is called a semimodular. If moreover,

(1″) $\rho(f) = 0$ if and only if $f = 0$,

then ρ is called a modular. If ρ is a pseudomodular in $\mathfrak{X}$ then the set defined by

$$\mathfrak{X}_\rho \;=\; \{f \in \mathfrak{X} \;;\; \lim_{\lambda \to 0}\; \rho(\lambda f) = 0\}$$

is called a modular space. $\mathfrak{X}_\rho$ is a vector subspace of $\mathfrak{X}$. For a pseudomodular ρ in $\mathfrak{X}$ we may define an F-seminorm by the formula:

$$\|f\|_\rho = \inf\Big\{u > 0 \;;\; \rho\Big(\tfrac{f}{u}\Big) \leq u\Big\}.$$

If ρ is an s-convex pseudomodular then the functional given by

$$\|f\|_\rho = \inf\Big\{u^s > 0 \;;\; \rho\Big(\tfrac{f}{u}\Big) \leq 1\Big\}$$

is an s-seminorm in $\mathfrak{X}_\rho$ (a seminorm for $s = 1$). Observe that the previous formulas define F-norms and s-norms respectively, if ρ is a modular or at least a semimodular. One can check that $\|f_k - f\|_\rho \to 0$ is equivalent to $\rho(\alpha(f_k - f)) \to 0$ for all $\alpha > 0$.

It is also an important fact that $\rho(f) \leq \|f\|_\rho$ provided $\|f\|_\rho < 1$.

We say that a sequence (f_k) is modular convergent (briefly: ρ-convergent) to $f \in \mathfrak{S}_\rho$ if there exists a $\lambda > 0$ such that $\rho(\lambda(f_k - f)) \to 0$ as $k \to \infty$. A modular ρ is called :

(a)   Right continuous, if $\lim\limits_{\lambda \to 1^+} \rho(\lambda f) = \rho(f)$ for all $f \in \mathfrak{S}_\rho$;

(b)   Left continuous, if $\lim\limits_{\lambda \to 1^-} \rho(\lambda f) = \rho(f)$ for all $f \in \mathfrak{S}_\rho$;

(c)   Continuous, if it is both right and left continuous.

In this way, the Orlicz space becomes a modular space, where $\mathfrak{S} = M(X,\mathbb{R})$ and the modular ρ is defined by

$$\rho(f) = \int_{\mathbb{R}} \phi\Big(|f(x)|\Big) \, dx.$$

On the base of the modular theory J. Musielak and W. Orlicz founded in 1959 a theory of Musielak-Orlicz spaces, i.e. the modular spaces induced by modulars of the following form:

$$\rho(f) = \int_X \phi\Big(x, |f(x)|\Big) \, d\mu(x),$$

where $\phi : X \times \mathbb{R}^+ \to \mathbb{R}^+$ is a function, continuous and increasing to infinity in the second variable, and is measurable in the first one (more precise definitions will be presented in Chapter 4). Such spaces have been studied for almost thirty years and there is known a large set of applications of such spaces in various parts of analysis. They were also generalized in many directions, e.g. some generalizations to the case of vector valued functions have been considered and many authors have investigated spaces generated by families of Musielak-Orlicz modulars. Such spaces have many applications in probability and mathematical statistics.

We may observe, however, the situation where on the one hand we have a very abstract general theory of modular spaces which cannot give proper answers to many interesting questions and, on the other hand, spaces constructed on the image of Musielak-Orlicz spaces. In the latter case, the concepts from Musielak-Orlicz theory do not fit the new demands. Another common difficulty consists in the fact that the theory of Musielak-Orlicz spaces, though very useful, is not structural, in the sense that many operations like taking sums or passing to equivalent modulars lead beyond the class of Musielak-Orlicz spaces.

That is why the author tried to find a theory situated somewhere in between, i.e. to find a class of modular spaces given by modulars or semimodulars not of any particular forms but, nevertheless, having much more convenient properties than the abstract modulars can possess. In other words, the aim of our theory is to give a useful tool for applications whenever there is a need to introduce a function space by means of functionals which have some reasonable properties but which are far from being norms, F-norms or any of the classical modulars. On the other hand, it is clear that this class of function spaces should contain many well known modular spaces of functions.

The main idea of our theory consists in considering function semimodulars which are functionals both of functions and sets; this enables us to proceed in a similar way as in the case of semimodulars given by integrals, which are elementary examples of functionals depending both on functions and sets.

The material is divided into seven chapters. At the end of each there is a short section of bibliographical remarks. The notes refer to the list of references at the end of the book.

In Chapters 2 and 3 the basic theory of modular function spaces is presented. Next, Chapter 4 contains special cases and examples of function semimodulars and function modulars; the case of nonmonotone functionals with similar properties to function modulars is considered as well as the question when they can be really regarded as function modulars. Chapter 5 is devoted to some constructions involving function modulars; we define countably and uniformly countably modulared function spaces and present some applications to the theory of summation of integrals. In particular, we are interested in the following problem: when a function f which is summable with respect to every method from a countable family of methods is uniformly summable . The results of this chapter are then used in the next one, while considering the nonlinear operators. In Chapter 6 we deal, as it was said, with nonlinear operators. We consider a problem of finding a maximal domain of continuity for such operators. This interesting problem may itself serve as a justification of the theory of modular function spaces since, surprisingly, we can obtain strong results in a very general situation, when no other theory can help. Next we consider the problem of existence of fixed points for ρ-nonexpansive and ρ-contractive operators. Chapter 7 presents some applications to the problems of analytic extension of functions of several complex variables. The surprising interaction between the theory of modular function spaces, complex analysis and approximation theory results in some conditions, both necessary and sufficient for analytic extension, expressed in terms of approximation by polynomials.

Bibliographical remarks

The monographic exposition of the theory of Orlicz spaces may be found in the book of Krasnosel'skii and Rutickii [1]. For a current review of the theory of Musielak-Orlicz spaces and modular spaces the reader is referred to the book of Musielak [1]. The results of the Leiden school in Banach function spaces were published first in Luxemburg's thesis [1] and then in a series of papers of Luxemburg and Zaanen in years 1955 - 1966 (Luxemburg [2] - [4], Luxemburg and Zaanen [1] - [13]). The book of Musielak contains most references on generalizations of Musielak-Orlicz spaces. As regards spaces of vector valued functions, we want to mention basic expository works of Kozek [1], [2] and Turett [2]. Spaces induced by families of modulars were first considered by Musielak and Waszak [2], [3] in 1970, Rosenberg [1] in 1970 and, in a more general setting by Kaminska [1] and by Drewnowski and Kaminska [1]. The concept of modular function spaces was introduced by Kozlowski in his papers [4], [5].

2. Function Semimodulars

2.1 BASIC DEFINITIONS

Let X be a nonempty set and $\mathcal{P}$ be a nontrivial δ-ring of subsets of X, i.e. the ring closed with respect to the forming of countable intersections. Recall that a nonempty class $\mathcal{P}$ of sets is called a ring if $\mathcal{P}$ is closed under the formation of finite unions and differences. Let Σ be the smallest σ-algebra of subsets of X such that Σ contains $\mathcal{P}$ and

(i) $\mathcal{P}$ is and ideal in Σ, i.e. $E \cap A \in \mathcal{P}$ for every $E \in \mathcal{P}$, $A \in \Sigma$;

(ii)   There exists a sequence $X_i \in \mathcal{P}$ such that $X_i \uparrow X$, i.e. X is a union of X_i and (X_i) is nondecreasing.

We do not have any measure here but, roughly speaking, the δ-ring $\mathcal{P}$ will play the role of the δ-ring of finite measure sets. If $E \subset X$ then 1_E will stand for its characteristic function.

11

$(S, |\cdot|)$ will always denote a Banach space which will be a range space for functions acting from X. By a $\mathcal{P}$-simple function on X with values in a Banach space S we mean a function of the form

$$g = \sum_{i=1}^{n} r_i 1_{E_i}, \text{ where } r_i \in S, \ E_i \in \mathcal{P}, \ E_i \cap E_j = \emptyset \text{ for } i \neq j.$$

The linear space of all $\mathcal{P}$-simple functions will be denoted by $\mathcal{E}$.

A function $f : X \rightarrow S$ is called measurable if there is a sequence of $\mathcal{P}$-simple functions (f_n) such that $f_n(x) \rightarrow f(x)$ for every $x \in X$. It is well known that a function f is measurable if and only if it is separably valued in S and for every open set $B \subset S$ there holds $f^{-1}(B) \in \Sigma$ (cf. Dunford and Schwartz [1], III.6.9). Thus, if f is measurable then $|f(x)|$ is a scalar measurable function of a variable $x \in X$. It follows from the properties of scalar measurable functions that to every measurable f there corresponds a sequence of $\mathcal{P}$-simple functions (f_n) such that $|f_n(x)| \uparrow |f(x)|$ for every $x \in X$. The vector space consisting of all measurable functions acting from X into S will be denoted by M(X,S). By the support of a measurable function f we will mean the measurable set

$$\mathrm{supp}(f) = \{ x \in X ; \ f(x) \neq 0 \}.$$

An operator $T : D \rightarrow H$, where $D \subset M(X,S)$ and H is an arbitrary vector space, will be called additive (some authors use names: orthogonally or disjointly additive operators) if T(f + g) = T(f) + T(g) whenever supports of these functions are disjoint.

A set function $\mu : \Sigma \to [0,\infty]$ is called a subadditive measure on Σ if:

(1) $\mu(\emptyset) = 0$;

(2) $\mu(A \cup B) \leq \mu(A) + \mu(B)$
       whenever A, B $\in \Sigma$, A $\cap$ B = $\emptyset$;

(3) $\mu(A) \leq \mu(B)$ if A, B $\in \Sigma$ and A $\subset$ B.

If, moreover,

$$\mu \left(\bigcup_{n=1}^{\infty} E_n \right) \leq \sum_{n=1}^{\infty} \mu(E_n)$$

for every sequence of sets $E_n \in \Sigma$, then μ will be called a σ-subadditive measure.

A subadditive measure is said to be order continuous if $\mu(E_n) \to 0$ for every sequence $(E_n) \subset \Sigma$ such that $E_n \downarrow \emptyset$. Furthermore, μ is called exhaustive if $\mu(E_n) \to 0$ whenever $(E_n) \subset \Sigma$ are mutually disjoint. It was proved by Drewnowski in [2], Cor. 5.4, that if μ is a subadditive measure defined on a σ-ring then μ is order continuous if and only if μ is both exhaustive and σ-subadditive.

2.1.1 DEFINITION. A functional $\rho : \mathcal{E} \times \Sigma \rightarrow [0,\infty]$ is called a function semimodular if and only if the following conditions are satisfied:

P1. $\rho(0,E) = 0$ for every $E \in \Sigma$;

P2. $\rho(f,E) \leq \rho(g,E)$ whenever $|f(x)| \leq |g(x)|$ for all $x \in E$ and all f, g $\in \mathcal{E}$ ($E \in \Sigma$);

P3. $\rho(f,\cdot) : \Sigma \rightarrow [0,\infty]$ is a σ-subadditive measure for every $f \in \mathcal{E}$;

P4. $\rho(\alpha,A) \rightarrow 0$ as $\alpha \downarrow 0$ for every $A \in \mathcal{P}$, where for the sake of simplicity we denote
$\rho(\alpha,A) = \sup\{\rho(r1_A,A) ; r \in S, |r| \leq \alpha \}$;

P5.   There exists $\alpha_0 \geq 0$ such that $\sup\limits_{\beta>0} \rho(\beta,A) = 0$ whenever $\sup\limits_{\alpha>\alpha_0} \rho(\alpha,A) = 0$;

P6. $\rho(\alpha,\cdot)$ is order continuous on $\mathcal{P}$ (for every $\alpha > 0$).

The definition of ρ is then extended to all f in $M(X,S)$ by defining that

$\rho(f,E) = \sup\{\rho(g,E) ; g \in \mathcal{E}, |g(x)| \leq |f(x)| $ for each $x \in E\}$.

In this sense we will also understand the notation $\rho(\alpha,E)$ for sets E not belonging to $\mathcal{P}$. For the sake of simplicity, we shall write $\rho(f)$ instead of $\rho(f,X)$. Let us observe that, for the time being, the notion of function semimodular has nothing to do with the definition of semimodular given in the previous chapter. The justification of this terminology will be given in Theorem 2.1.4. The following properties of function semimodulars are the immediate consequence of the definition.

2.1.2 PROPOSITION. For f and g in M(X,S), and E $\in$ Σ:

(a) $\rho(f,E) \leq \rho(g,E)$ if $|f(x)| \leq |g(x)|$ for all $x \in E$;

(b) $\rho(f,\cdot) : \Sigma \to [0,\infty]$ is a σ-subadditive measure for every f$\in$M(X,S);

(c) $\rho(f,E) = \rho(g,E)$ whenever $f(x) = g(x)$ for $x \in E$;

(d) $\rho(f,E) \leq \rho(g,E)$ if $|f(x)| \leq |g(x)|$ for all $x \in X$;

(e) $\rho(f,E) = 0$ if $f(x) = 0$ for every $x \in E$;

(f) $\rho(f,E) = \rho\Big(f,\, E \cap \text{supp}(f)\Big)$;

(g) $\rho(f,E) = \rho(f1_E,E)$.

We will introduce now a basic notion of ρ-null sets and the relation of being equal ρ almost everywhere, which play a similar role as sets of measure zero and equality almost everywhere in the measure sense do for the theory of L^p-spaces or Orlicz spaces.

2.1.3 DEFINITION. A set $E \in \Sigma$ is said to be ρ-null if and only if $\rho(\alpha,E) = 0$ for every $\alpha > 0$. A property $w(x)$ is said to hold ρ almost everywhere (ρ-a.e.) if the exceptional set of x in X such that $w(x)$ does not hold is ρ-null.

Let us observe that a countable union of ρ-null sets is still ρ-null. In view of the property P5, if $\rho(\alpha,E) = 0$ for an $\alpha > \alpha_0$ then E is ρ-null. In the sequel, we will identify sets A and B whose symmetric difference $A \triangle B$ is ρ-null; similarly, we will identify the functions which differ on ρ-null sets only. Now, we are ready to prove that ρ is a semimodular in a strict sense.

2.1.4 THEOREM. The functional $\rho : M(X,S) \rightarrow [0,\infty]$ is a semimodular.

PROOF. Let us recall that by the definition $\rho(f) = \rho(f,X)$ and, once again, that we identify functions which are equal ρ-a.e. It is clear that if $f = 0$ ρ-a.e., then $\rho(f) = 0$. Let us assume now that $\rho(\lambda f) = 0$ for every $\lambda > 0$. Take any sequence of simple functions $g_n \in \mathcal{S}$ such that $|g_n| \uparrow |f|$ ρ-a.e. and observe that

$$\text{supp}(f) = \bigcup_{n=1}^{\infty} \text{supp}(g_n).$$

Since $\rho(\lambda g_n) \leq \rho(\lambda f) = 0$ for every $\lambda > 0$, it follows that

$$\rho\Big(\alpha_n, \text{supp }(g_n)\Big) = 0 \quad \text{for an } \alpha_n > \alpha_0,$$

which implies, in view of P5, that $\text{supp}(g_n)$ is a ρ-null (for every $n \in \mathbb{N}$) and, consequently, $\text{supp}(f)$ is ρ-null, i.e. $f = 0$ ρ-a.e. Since (2) from the definition of the semimodular is clearly satisfied, it remains to prove (3). Let f, g $\in$ M(X,S) and $0 \leq \alpha$, $\beta \leq 1$ be such that $\alpha + \beta = 1$. Compute

$$|\alpha f(x) + \beta g(x)| \leq \alpha|f(x)| + \beta|g(x)| \leq \max\{|f(x)|, |g(x)|\}.$$

Put

$$E_1 = \{x \in X ; |f(x)| \geq |g(x)|\},$$

$$E_2 = X \setminus E_1, \quad h = f1_{E_1} + g1_{E_2}.$$

Then

$$|\alpha f(x) + \beta g(x)| \leq |h(x)| \quad \text{for every } x \in X$$

and

$$\rho(\alpha f + \beta g) \leq \rho(h) \leq \rho(h, E_1) + \rho(h, E_2)$$

$$= \rho(f, E_1) + \rho(g, E_2) \leq \rho(f) + \rho(g).$$

This completes the proof.

If α_0 from P5 is equal to zero then ρ is a modular and will be called a function modular. In the case ρ satisfies all properties from Definition 2.1 except P5, then ρ is a pseudomodular and will be called a function pseudomodular. According to the general modular theory we can define now a modular function space induced by the function modular ρ, i.e. the space defined by the formula

$$L_\rho = \{\ f \in\ M(X,S)\ ;\ \rho(\lambda f) \rightarrow 0\ \text{as}\ \lambda \rightarrow 0^+\ \}.$$

If ρ is a function semimodular or function modular then we may equip L_ρ with an F-norm (s-norm or norm in case of s-convex or respectively convex function semimodulars), which will always be denoted by $\|\cdot\|_\rho$. In general, our results are not affected by convexity of the semimodular and, therefore, if not necessary we will not distinguish between these cases or between F-norms, s-norms and norms given by them.

It is evident in view of P4 that L_ρ contains all bounded functions with supports from $\mathcal{P}$; in particular, $\mathcal{E} \subset L_\rho$. Therefore, P4 guarantees that L_ρ is nontrivial. The fact that L_ρ contains all bounded functions with supports from $\mathcal{P}$ seems to be essential for the entire theory and its applications. By the same P4 we get immediately another important fact.

2.1.5 PROPOSITION. If $(f_n) \subset M(X,S)$ converges uniformly to a function $f \in M(X,S)$ on a set $E \in \mathcal{P}$ then $\rho(\alpha(f_n-f), E) \rightarrow 0$ for all $\alpha > 0$, i.e. $\|(f_n-f)1_E\|_\rho \rightarrow 0$.

In other words, Proposition 2.1.5 states that uniform convergence on sets from $\mathcal{P}$ is stronger than the convergence with respect to the norm $\|\cdot\|_\rho$.

2.2 FATOU PROPERTY

On account of the obvious analogy with the classical case we make the following definition.

2.2.1 DEFINITION. A function semimodular is said to have the Fatou property if and only if $\rho(f_n) \uparrow \rho(f)$ whenever $|f_n| \uparrow |f|$ ρ-a.e.

2.2.2 THEOREM. The following conditions are equivalent:

(i) ρ has the Fatou property;

(ii) ρ is a left continuous semimodular.

PROOF. (i) $\Rightarrow$ (ii) Evident.

(ii) $\Rightarrow$ (i) Suppose $|f_n| \uparrow |f|$ ρ-a.e. and denote $\gamma = \lim\limits_{n \to \infty} \rho(f_n)$. Since $(\rho(f_n))$ is a nondecreasing sequence we conclude that γ is well defined and for all natural n there holds $\rho(f_n) \leq \gamma \leq \rho(f)$. Thus, it suffices to prove that $\gamma \geq \rho(f)$. Let $H \in \Sigma$ be a ρ-null exceptional set, $H = \{x \in X ; f_n(x)$ does not converge to $f(x)\}$, and let $W = X \setminus H$. Let us choose a $\mathcal{P}$-simple function g such that $|g(x)| \leq |f(x)|$ for every $x \in X$ and fix a number $\lambda \in (0,1)$. Let

$$E_n = \{ x \in W ; \lambda|g(x)| \leq |f_n(x)| \} \text{ and } D_n = W \setminus E_n.$$

Observe that (D_n) is nonincreasing. It follows from the definition of D_n that

$$x \in \bigcap_{n=1}^{\infty} D_n$$

if and only if $x \in W$ and $|f_n(x)| < \lambda|g(x)| < |f(x)|$ for all $n \in \mathbb{N}$. Thus, $f_n(x)$ does not converge to $f(x)$ which contradicts the fact that $x \in W$. Hence,

$$\bigcap_{n=1}^{\infty} D_n = \emptyset$$

and consequently $D_n \downarrow \emptyset$. It is evident that $D_n \subset W \cap \text{supp}(g)$ and, since $\text{supp}(g) \in \mathcal{P}$, we get $D_n \in \mathcal{P}$. Fix $\epsilon > 0$, by P6 there exists a natural n_0 such that

$$\rho(\lambda g, D_n) < \epsilon \text{ for } n \geq n_0.$$

Thus,

$$\rho(\lambda g) \leq \rho(\lambda g, H) + \rho(\lambda g, E_n) + \rho(\lambda g, D_n)$$

$$\leq \rho(f_n, E_n) + \epsilon \leq \rho(f_n) + \epsilon \leq \gamma + \epsilon.$$

Since $\epsilon > 0$ was chosen arbitrarily, we conclude that $\rho(\lambda g) \leq \gamma$

which implies that $\rho(g) \leq \gamma$ because ρ is left continuous. The last inequality holds for arbitrary $g \in \mathcal{E}$ such that $|g(x)| \leq |f(x)|$, so we have finally $\rho(f) \leq \gamma$ and the proof is complete.

The next result describes the Fatou property in terms of $\|\cdot\|_\rho$.

2.2.3 THEOREM. If ρ has the Fatou property then $\|f_n\|_\rho \uparrow \|f\|_\rho$ whenever $|f_n| \uparrow |f|$ ρ-a.e.

PROOF. It follows from Proposition 2.1.2 (d) and from the definition of $\|\cdot\|_\rho$ that the sequence $(\|f_n\|_\rho)$ is nondecreasing. Denoting then

$$\gamma = \lim_{n \to \infty} \|f_n\|_\rho,$$

assume to the contrary that $\gamma + \epsilon < \|f\|_\rho$ for some $\epsilon > 0$. Let $\alpha_n > 0$ be such that

$$\rho\left(\frac{f_n}{\alpha_n}\right) \leq \alpha_n \quad \text{and} \quad \alpha_n < \gamma + \epsilon.$$

Such α_n exists because

$$\|f_n\|_\rho < \gamma + \epsilon \quad \text{for all n} \in \mathbb{N}.$$

Hence,

$$\rho\left(\frac{f_n}{\gamma+\epsilon}\right) \leq \rho\left(\frac{f_n}{\alpha_n}\right) \leq \alpha_n < \gamma + \epsilon.$$

Since ρ has the Fatou property it follows that

$$\rho\left(\frac{f_n}{\gamma+\epsilon}\right) \uparrow \rho\left(\frac{f}{\gamma+\epsilon}\right)$$

so that

$$\rho\left(\frac{f}{\gamma+\epsilon}\right) \leq \gamma + \epsilon$$

and finally

$$\|f\|_\rho \leq \gamma + \epsilon.$$

Contradiction.

2.2.4 COROLLARY. If ρ has the Fatou property then for every $f \in M(X,S)$ there holds

$$\|f\|_\rho = \sup \{ \|g\|_\rho \; ; g \in \mathcal{E}, |g(x)| \leq |f(x)| \text{ for every } x \in X \}.$$

2.3 COMPLETENESS OF MODULAR FUNCTION SPACES

We will now take under our consideration the question of completeness of modular function spaces. It will turn out that all modular fuction spaces are complete, i.e. they are always Frèchet or Banach spaces. Before we are able to prove the

Completeness Theorem, however, we have to develop further our basic theory of function semimodulars. Let us start with the following definition.

2.3.1 DEFINITION. Let f_n and f belong to $M(X,S)$. We say that (f_n) converges to f in submeasure (ρ_α) and write $f_n \to \hat{f} (\rho_\alpha)$ if and only if for every $\epsilon > 0$ there holds

$$\rho\Big(\alpha, \{ x \in X \, ; \, |f_n(x) - f(x)| \geq \epsilon \}\Big) \to 0 \ \text{ as } \ n \to \infty.$$

We say that (f_n) converges to f in the submeasure (ρ) and write $f_n \to f (\rho)$ whenever $f_n \to f (\rho_\alpha)$ for all $\alpha > 0$. The definition of Cauchy sequences is analogous. The reader should be aware of the difference between convergence in submeasure and convergence ρ almost everywhere. The relation between those notions will be disscussed in Proposition 2.3.5.

2.3.2 PROPOSITION. If (f_n) is a Cauchy sequence in L_ρ then (f_n) is a Cauchy sequence in submeasure (ρ).

PROOF. Let us fix $\epsilon > 0$, $\alpha > 0$ and denote

$$E_{n,k}(\epsilon) = \{ x \in X \, ; \, |f_n(x) - f_k(x)| \geq \epsilon \}.$$

Then

$$\rho(\epsilon, E_{n,k}(\epsilon)) \leq \rho(f_n - f_k, E_{n,k}(\epsilon)) \leq \rho(f_n - f_k).$$

Hence,

$$\rho(\alpha, E_{n,k}(\epsilon)) \le \rho\left(\tfrac{\alpha}{\epsilon}\,(f_n - f_k),\, E_{n,k}(\epsilon)\right) \le \rho\left(\tfrac{\alpha}{\epsilon}\,(f_n - f_k)\right) \to 0$$

as   n, k $\to \infty$.

Similarly, there holds the following result.

2.3.3 PROPOSITION. If f_n, f $\in L_\rho$ and $\|f_n - f\|_\rho \to 0$ then $f_n \to 0\ (\rho)$.

We want to prove now a version of the Egoroff Theorem.

2.3.4 THEOREM. Let f_n, f $\in$ M(X,S) and $f_n \to$ f ρ-a.e. There exists a nondecreasing sequence of sets $H_k \in \mathcal{P}$ such that $H_k \uparrow$ X and (f_n) converges uniformly to the function f on every H_k.

PROOF. Let us fix a positive number $\alpha > \alpha_0$. Let $X_n \uparrow$ X and $X_n \in \mathcal{P}$. Let us fix temporarily n $\in$ N. We shall prove that there exists a set $F_n \in \mathcal{P}$ such that $F_n \subset X_n$ and

$$\rho(\alpha, X_n \setminus F_n) < \tfrac{1}{2^n} \quad \text{and} \quad f_m \rightrightarrows \text{ f on } F_n.$$

Though this part of proof does not differ from the proof of classical theorem of Egoroff (see e.g. Halmos [1]), we present it in detail because of the fundamental role played by the Egoroff Theorem in the theory of modular function spaces.

By omitting, if necessary, a ρ-null set from X, we may assume that the sequence (f_n) converges to f everywhere. Put

$$E_k^m = \bigcap_{i=k}^{\infty} \left\{ x \in X_n \,;\, |f_i(x) - f(x)| < \tfrac{1}{m} \right\}$$

and observe that $E_1^m \subset E_2^m \subset \cdots$ and, since $f_n(x) \to f(x)$ for $x \in X_n$,

$$\bigcup_{k=1}^{\infty} E_k^m \supset X_n$$

for every $m \in \mathbb{N}$. Hence, there exists a positive integer $n_0 = n_0(m)$ such that

$$\rho\left(\alpha,\, X_n \setminus E_{n_0(m)}^m \right) < \frac{1}{2^{m+n}}.$$

Denoting

$$F_n = \bigcap_{m=1}^{\infty} E_{n_0(m)}^m$$

we have then

$$\rho(\alpha, X_n \setminus F_n) = \rho\left(\alpha,\, X_n \setminus \bigcap_{m=1}^{\infty} E_{n_0(m)}^m \right)$$

$$= \rho\left(\alpha,\, \bigcup_{m=1}^{\infty} (X_n \setminus E_{n_0(m)}^m) \right) \leq \sum_{m=1}^{\infty} \rho\left(\alpha,\, X_n \setminus E_{n_0(m)}^m \right) < \frac{1}{2^n}.$$

If $x \in F_n$ then to every $m \in \mathbb{N}$ there corresponds $k_m \in \mathbb{N}$ such that for $i \geq k_m$

$$|f_i(x) - f(x)| < \tfrac{1}{m},$$

i.e. (f_i) converges uniformly to the function f on F_n. Let us define

$$H_k = \bigcup_{n=1}^{k} F_n \quad \text{and} \quad H = \bigcup_{k=1}^{\infty} H_k.$$

Clearly, (H_k) is a nondecreasing sequence of sets from $\mathscr{P}$ such that

$$\rho(\alpha, X_k \setminus H) \leq \rho(\alpha, X_k \setminus H_k) \leq \rho(\alpha, X_k \setminus F_k) < \frac{1}{2^k}$$

and $f_m \rightrightarrows f$ on every H_k. It remains to prove that $X \setminus H$ is ρ-null. Indeed,

$$\rho(\alpha, X \setminus H) \leq \rho(\alpha, X_k \setminus H) + \sum_{n=k+1}^{\infty} \rho(\alpha, X_n \setminus H)$$

$$< \frac{1}{2^k} + \sum_{n=k+1}^{\infty} \frac{1}{2^n},$$

for every $k \in \mathbb{N}$. Thus $\rho(\alpha, X \setminus H) = 0$.

The next Proposition is also similar to classical results from measure theory. We give the proofs in order to present a self-contained discussion of the basic facts of our theory.

2.3.5 PROPOSITION. Let a sequence (f_n) satisfy the Cauchy condition in submeasure (ρ_α); then there are $f \in M(X,S)$ and a subsequence (g_n) of (f_n) such that:

(i) $f_n \to f \ (\rho_\alpha)$;

(ii) $g_n \to f \ \rho$-a.e.

PROOF. (ii) For any $k \in \mathbb{N}$ we can find an integer $n_0(k)$ such that if $n \geq n_0(k)$ and $m \geq n_0(k)$, then

$$\rho\left(\alpha, \left\{x \in X \ ; \ |f_n(x) - f_m(x)| \geq \tfrac{1}{2^k}\right\}\right) < \tfrac{1}{2^k}.$$

Taking $n_1 = n_0(1)$, $n_k = \max \ (n_{k-1}, \ n_0(k))$ we get the subsequence $g_k = f_{n_k}$. Define

$$E_k = \left\{x \in X \ ; \ |g_k(x) - g_{k+1}(x)| \geq \tfrac{1}{2^k}\right\}.$$

For every $j \geq i \geq k$ and every $x \in F_k = X \setminus \bigcup_{n=k}^{\infty} E_n$ we have

$$|g_i(x) - g_j(x)| \leq \sum_{m=i}^{\infty} |g_m(x) - g_{m+1}(x)| < \tfrac{1}{2^{i-1}}.$$

Compute

$$\rho(\alpha, X \setminus F_k) = \rho\left(\alpha, \bigcup_{n=k}^{\infty} E_n\right)$$

$$\leq \sum_{n=k}^{\infty} \rho(\alpha, E_n) \leq \sum_{n=k}^{\infty} \tfrac{1}{2^n} = \tfrac{1}{2^{k-1}}.$$

Let F be an union of all sets F_k; then $\rho(\alpha, X \setminus F) = 0$ and consequently $X \setminus F$ is ρ-null. Since S is complete, for every

$x \in F$ there exists $f(x) \in S$ such that $g_n(x) \rightarrow f(x)$.

(i) By (ii) we can choose a subsequence (g_k) which is a Cauchy sequence in the sense of convergence ρ-a.e. We write $f(x) = \lim\limits_{k \to \infty} g_k(x)$ for every $x \in X$ for which the limit exists. We observe that, for every $\epsilon > 0$,

$$\{x \in X \; ; \; |f_n(x) - f(x)| \geq \epsilon \}$$

$$\subset \left\{x \in X \; ; \; |f_n(x) - g_k(x)| \geq \tfrac{\epsilon}{2}\right\} \cup \left\{x \in X \; ; \; |g_k(x) - f(x)| \geq \tfrac{\epsilon}{2}\right\}.$$

Hence,

$$\rho\left(\alpha, \{x \in X \; ; \; |f_n(x) - f(x)| \geq \epsilon\}\right)$$

$$\leq \rho\left(\alpha, \left\{x \in X \; ; \; |f_n(x) - g_k(x)| \geq \tfrac{\epsilon}{2}\right\}\right)$$

$$+ \rho\left(\alpha, \left\{x \in X \; ; \; |g_k(x) - f(x)| \geq \tfrac{\epsilon}{2}\right\}.$$

The first term on the right is by hypothesis arbitrarily small if n and k are sufficiently large. Proving (ii) we observed that for $x \in F_k$

$$|g_i(x) - g_j(x)| < \frac{1}{2^{i-1}}$$

for $j \geq i \geq k$. Hence,

$$|g_k(x) - f(x)| \leq \frac{1}{2^{k-1}},$$

because $g_j(x) \to f(x)$ as $j \to \infty$. Thus, taking k large enough, we have

$$\left\{ x \in X \; ; \; |g_k(x) - f(x)| \geq \tfrac{\epsilon}{2} \right\} \cap F_k = \emptyset.$$

Then

$$\rho\left(\alpha, \left\{ x \in X \; ; \; |g_k(x) - f(x)| \geq \tfrac{\epsilon}{2} \right\} \right)$$

$$\leq \rho\left(\alpha, F_k \cap \left\{ x \in X \; ; \; |g_k(x) - f(x)| \geq \tfrac{\epsilon}{2} \right\} \right)$$

$$+ \rho\left(\alpha, X \setminus F_k \right) \leq 0 + \frac{1}{2^{k-1}} \to 0.$$

Finally, $\rho(\alpha, \{ x \in X \; ; \; |f_n(x) - f(x)| \geq \epsilon \}) \to 0$ as $n \to \infty$ and (i) has been proved.

2.3.6 PROPOSITION. If (f_k) is a Cauchy sequence in (ρ) then there exists a function $f \in M(X,S)$ such that $f_k \to f \, (\rho)$.

PROOF. In virtue of the previous result, to every $\alpha > 0$ there exists a function $h_\alpha \in M(X,S)$ such that $f_k \to h_\alpha \, (\rho_\alpha)$. Let us choose an arbitrary sequence of scalars $\alpha_0 < \alpha_1 < \cdots$, $\alpha_n \to \infty$ and define $g_n = h_{\alpha_n}$. If $i \geq j$ then $f_k \to g_i \, (\rho_{\alpha_j})$ because

$$f_k \to g_i \, (\rho_{\alpha_i}) \text{ and } \rho(\alpha_j, \cdot) \leq \rho(\alpha_i, \cdot).$$

Since $f_k \to g_j \, (\rho_{\alpha_j})$, it follows easily that $g_i = g_j$ ρ-a.e. Put $f = g_0$ and take an $\alpha > 0$. For a certain natural n we have

$\alpha \le \alpha_n$. For a given $\epsilon > 0$ there holds then

$$\rho\Big(\alpha, \{x \in X \; ; \; |f_k(x) - f(x)| \ge \epsilon\}\Big)$$

$$\le \rho\Big(\alpha_n, \{x \in X \; ; \; |f_k(x) - f(x)| \ge \epsilon\}\Big)$$

$$= \rho\Big(\alpha_n, \{x \in X \; ; \; |f_k(x) - g_n(x)| \ge \epsilon\}\Big) \to 0$$

as $k \to \infty$. Finally, $f_k \to f \; (\rho_\alpha)$ and then $f_k \to f \; (\rho)$ because α was fixed arbitrarily.

Propositions 2.3.2 and 2.3.6 are the main tools for proving the basic result of this section, the theorem on completeness of L_ρ.

2.3.7 THEOREM. The linear metric space L_ρ is complete.

PROOF. Let (f_n) be a Cauchy sequence in the space $(L_\rho, \|\cdot\|_\rho)$. By Proposition 2.3.2 the sequence (f_n) is also a Cauchy sequence in the sense of convergence in submeasure. Hence, in virtue of Proposition 2.3.6 there exists a function $f \in M(X,S)$ such that $f_n \to f \; (\rho)$. Let us fix a number $\alpha > 0$; choose then a subsequence (g_k) of (f_n) such that

$$\rho(2\alpha(g_k - g_{k+n})) < \frac{1}{2^{k+1}} \text{ for } n \in \mathbb{N}.$$

Let us fix temporarily a natural number k and take arbitrary h ∈ 𝔖 such that

$$|h(x)| \leq |f(x) - g_k(x)| \text{ for all } x \in X.$$

Let us denote

$$a = \inf_{x \in X} |h(x)|, \quad b = \sup_{x \in X} |h(x)|,$$

$$A_n = \left\{ x \in X; |g_{k+n}(x) - f(x)| > \frac{a}{2} \right\}.$$

Since $g_n \rightarrow f (\rho)$ it follows that

$$\rho(\alpha b, A_n) < \frac{1}{2^{k+1}}$$

for n sufficiently large. On the other hand, for each $x \in X \setminus A_n$ we have

$$|h(x)| \leq |f(x) - g_k(x)| \leq |f(x) - g_{k+n}(x)| + |g_{k+n}(x) - g_k(x)|$$

$$\leq \frac{a}{2} + |g_{k+n}(x) - g_k(x)| \leq \frac{|h(x)|}{2} + |g_{k+n}(x) - g_k(x)|.$$

Hence,

$$|h(x)| \leq 2 |g_{k+n}(x) - g_k(x)|$$

for $x \in X \setminus A_n$ and finally for n sufficiently large there holds

$$\rho(\alpha h) \leq \rho(\alpha h, A_n) + \rho(\alpha h, X \setminus A_n)$$

$$\leq \rho(\alpha b, A_n) + \rho(2\alpha|g_{k+n} - g_k|, X \setminus A_n)$$

$$\leq \rho(\alpha b, A_n) + \rho\Big(2\alpha(g_{k+n} - g_k)\Big)$$

$$\leq \frac{1}{2^{k+1}} + \frac{1}{2^{k+1}} = \frac{1}{2^k}.$$

Since h was an arbitrarily chosen simple function such that

$$|h| \leq |f - g_k|,$$

it follows that

$$\rho\Big(\alpha(f - g_k)\Big) < \frac{1}{2^k} \to 0 \text{ as } k \to \infty,$$

and therefore,

$$\|g_k - f\|_\rho \to 0.$$

We conclude then that $\|f_k - f\|_\rho \to 0$, because (f_n) is a Cauchy sequence in L_ρ.

We have to show now that $f \in L_\rho$. Let $\lambda_n \geq 0$, $\lambda_n \to 0$ and let $\lambda > 0$ be such that $\lambda_n \leq \lambda$ for n sufficiently large. Fix $\epsilon > 0$ and take $k \in \mathbb{N}$ such that

$$\rho\Big(\lambda(f_k - f)\Big) < \epsilon.$$

For n sufficiently large we obtain

$$\rho\Big(\lambda_n(f_k - f)\Big) \leq \rho\Big(\lambda(f_k - f)\Big) < \epsilon.$$

This inequality implies that

$$\rho\Big(\lambda_n(f_k - f)\Big) \to 0 \quad \text{as } n \to \infty.$$

Thus, $f_k - f$ belongs to L_ρ and finally $f \in L_\rho$ because $f_k \in L_\rho$ and L_ρ is a linear space. This completes the proof.

2.4 SUBSPACE E_ρ

Let us recall that a measurable function is said to have an absolutely continuous F-norm if and only if for every sequence $E_n \downarrow \emptyset$ there holds $\|f1_{E_n}\|_\rho \to 0$. We know that all bounded functions with supports from $\mathcal{P}$ have absolutely continuous F-norms $\|\cdot\|_\rho$ but, generally speaking, not all members of L_ρ have this property. We shall distinguish, therefore, the class of all functions with absolutely continuous F-norms. Since the condition $\|f1_{E_n}\|_\rho \to 0$ may be equivalently stated in the modular form, we make the following definition.

2.4.1 DEFINITION. Let E_ρ be a space of $f \in M(X,S)$ such that $\rho(\alpha f, \cdot)$ is order continuous for every $\alpha > 0$.

The position of E_ρ with respect to L_ρ is described in the next result.

2.4.2 THEOREM. E_ρ is a closed subspace of L_ρ.

PROOF. First we will prove that E_ρ is a linear subspace of L_ρ. Clearly E_ρ is a linear space; it suffices, therefore, to prove that $E_\rho \subset L_\rho$. Let $f \in E_\rho$ and $0 \le \lambda_n \to 0$. Since $f \in M(X,S)$ it follows that there exists a sequence $s_m \in \mathcal{S}$ such that $|s_m| \uparrow |f|$. By the Egoroff Theorem we get a sequence (H_i) such that $H_i \in \mathcal{P}$, $H_i \uparrow X$ and (s_m) converges uniformly on every H_k. Let us choose a $\lambda > 0$ such that $\lambda_n \le \lambda$ for all natural numbers n and fix an arbitrary number $\epsilon > 0$. Since $f \in E_\rho$, there exists an index k_0 such that

$$\rho\left(\lambda f, X \setminus H_{k_0}\right) < \tfrac{\epsilon}{3}.$$

In view of the uniform convergence of (s_m) on H_{k_0} we may choose a natural number m_0 such that

$$\rho\left(2\lambda(s_{m_0} - f), H_{k_0}\right) < \tfrac{\epsilon}{3}.$$

The function s_{m_0} belongs to $\mathcal{S} \subset L_\rho$ and thus, there exists an n_0 such that

$$\rho\left(2\lambda_n s_{m_0}, H_{k_0}\right) < \tfrac{\epsilon}{3} \quad \text{for } n \ge n_0.$$

Finally, for $n \ge n_0$ we have

$$\rho(\lambda_n f) \le \rho\left(\lambda_n f, X \setminus H_{k_0}\right) + \rho\left(\lambda_n f, H_{k_0}\right)$$

$$\leq \rho\Big(\lambda f, X \setminus H_{k_0}\Big) + \rho\Big(2\lambda(s_{m_0} - f), H_{k_0}\Big) + \rho\Big(2\lambda_n s_{m_0}, H_{k_0}\Big)$$

$$\leq \tfrac{\epsilon}{3} + \tfrac{\epsilon}{3} + \tfrac{\epsilon}{3} = \epsilon.$$

This means that the function f is a member of the space L_ρ. We will prove now that E_ρ is closed. Let $f_n \in E_\rho$, $\|f_n - f\|_\rho \to 0$, $f \in L_\rho$. Take a sequence of sets $E_n \in \Sigma$, $E_n \downarrow \emptyset$. Fix positive ϵ and α. We get

$$\rho\Big(2\alpha(f - f_{k_0})\Big) < \tfrac{\epsilon}{2}$$

for a certain index k_0. Since $f_{k_0} \in E_\rho$, it follows that for n sufficiently large

$$\rho\Big(2\alpha f_{k_0}, E_n\Big) < \tfrac{\epsilon}{2}.$$

Thus, for n sufficiently large there holds

$$\rho(\alpha f, E_n) \leq \rho\Big(2\alpha(f - f_{k_0}), E_n\Big) + \rho\Big(2\alpha f_{k_0}, E_n\Big)$$

$$\leq \rho\Big(2\alpha(f - f_{k_0})\Big) + \rho\Big(2\alpha f_{k_0}, E_n\Big) \leq \tfrac{\epsilon}{2} + \tfrac{\epsilon}{2} = \epsilon,$$

which implies that $f \in E_\rho$. Hence, E_ρ is a closed subspace of L_ρ.

The subspace E_ρ plays an important role in the theory, mainly because it possesses some convenient properties which are similar to the properties of the spaces of summable functions. The exposition of some basic properties of E_ρ will be started with the result which is called the Vitali Convergence Theorem on account of its clear similarity to the classic theorem of Vitali.

2.4.3 THEOREM. Let $f_n \in E_\rho$, $f \in L_\rho$ and $f_n \to f$ ρ-a.e.; then the following conditions are equivalent:

(i) $f \in E_\rho$ and $\|f_n - f\|_\rho \to 0$;

(ii) for every $\alpha > 0$ the subadditive measures $\rho(\alpha f_n, \cdot)$ are order equicontinuous, i.e. if $E_k \in \Sigma$, $E_k \downarrow \emptyset$ then

$$\sup_{n \in \mathbb{N}} \rho(\alpha f_n, E_k) \to 0 \quad \text{as } k \to \infty.$$

PROOF. (i) $\Rightarrow$ (ii)   Let us choose a sequence of sets $E_k \downarrow \emptyset$ from the σ-algebra Σ and fix arbitrarily positive ϵ and α. Since $\|f_n - f\|_\rho \to 0$, it follows that there exists $n_0 \in \mathbb{N}$ such that

$$\rho\Big(2\alpha(f_n - f)\Big) < \tfrac{\epsilon}{2} \quad \text{for } n \geq n_0.$$

Let k_0 be a natural number with

$$(2.4.4) \quad \rho(\alpha f_n, E_k) < \epsilon \quad \text{for } k \geq k_0 \text{ and } n = 1, 2, \dots, n_0 - 1.$$

Similarly, since $f \in E_\rho$, there exists $k_1 > k_0$ such that for $k \geq k_1$

$$\rho(2\alpha f, E_k) < \tfrac{\epsilon}{2}.$$

Thus,

$$(2.4.5) \qquad \rho(\alpha f_n, E_k) \leq \rho\Big(2\alpha(f_n - f)\Big) + \rho(2\alpha f, E_k) < \epsilon$$

for $k \geq k_1$ and $n \geq n_0$. The inequalities (2.4.4) and (2.4.5) give the desired result.

(ii) $\Rightarrow$ (i) Let us fix two positive numbers ϵ and α. Since $f_n \to f$ ρ-a.e. it follows by the Egoroff Theorem that there exists a sequence of sets $H_k \in \mathcal{P}$ such that $H_k \uparrow X$ and f_n uniformly converges to f on every H_k. By the order equicontinuity of semimodulars we can pick up an index k_0 such that

$$\rho\Big(2\alpha f_n, X \setminus H_{k_0}\Big) < \tfrac{\epsilon}{4} \quad \text{for } n \in \mathbb{N}.$$

Since f_n converges uniformly to f on the set $H_{k_0} \in \mathcal{P}$, one can find $n_0 \in \mathbb{N}$ such that

$$\rho\Big(2\alpha(f_n - f)\Big) < \tfrac{\epsilon}{4} \quad \text{for } n \geq n_0.$$

Let $n \geq n_0$ and $m \geq n_0$. We compute

$$\rho\Big(\alpha(f_n - f_m)\Big)$$

$$\leq \rho\Big(\alpha(f_n - f_m), X \setminus H_{k_0}\Big) + \rho\Big(\alpha(f_n - f_m), H_{k_0}\Big)$$

$$\leq \rho\Big(2\alpha f_n, X \setminus H_{k_0}\Big) + \rho\Big(2\alpha f_m, X \setminus H_{k_0}\Big)$$

$$+\rho\Big(2\alpha(f_n - f)\Big) + \rho\Big(2\alpha(f_m - f)\Big)$$

$$< \tfrac{\epsilon}{4} + \tfrac{\epsilon}{4} + \tfrac{\epsilon}{4} + \tfrac{\epsilon}{4} = \epsilon.$$

Hence, the sequence (f_n) satisfies the Cauchy condition in the sense of F-norm $\|\cdot\|_\rho$. Since $f_n \in E_\rho$ and E_ρ is complete as a closed subspace of the complete metric space L_ρ it follows that there exists a function $g \in E_\rho$ such that $\|f_n - g\|_\rho \to 0$. Therefore, in view of Proposition 2.3.5, the sequence (f_n) converges to g in submeasure. It follows again from Proposition 2.3.5 that there exists a subsequence of (f_n) which converges to g ρ-a.e. On the other hand, $f_n \to f$ ρ-a.e. Thus, $f = g$ ρ-a.e. and finally $\|f_n - f\|_\rho \to 0$, which is the desired result.

Making use of the Vitali Theorem one can easily prove the following result.

2.4.6 COROLLARY. For functions $f_n, f \in E_\rho$ the following two statements are equivalent:

(i) $\|f_n - f\|_\rho \to 0$;

(ii) $f_n \to f$ (ρ) and $\rho(\alpha f_n, \cdot)$ are order equicontinuous for every $\alpha > 0$.

As an immediate consequence of the Vitali Theorem we get the result of a great importance for applications, namely the following version of the Lebesgue Dominated Convergence Theorem.

2.4.7 THEOREM. If f_n, $f \in M(X,S)$, $f_n \to f$ ρ-a.e. and there exists a function $g \in E_\rho$ such that for every $n \in \mathbb{N}$ there holds $|f_n(x)| \leq |g(x)|$ ρ-a.e., then $\|f_n - f\|_\rho \to 0$.

This theorem will be now used to determine the space E_ρ.

2.4.8 THEOREM. The space E_ρ is the closure (in the sense of $\|\cdot\|_\rho$) of the space of all $\mathcal{P}$-simple functions.

PROOF. We will prove first that $cl_{\|\cdot\|_\rho}(\mathcal{E}) \subset E_\rho$. Take a function $g \in \mathcal{E}$, a number $\lambda > 0$ and a sequence $E_n \downarrow \emptyset$, $E_n \in \Sigma$. Denote

$$\alpha = \sup_{x \in X} |\lambda g(x)|,$$

and observe that

$$\rho(\lambda g, E_n) = \rho(\lambda g, \text{supp}(g) \cap E_n) \leq \rho(\alpha, \text{supp}(g) \cap E_n) \to 0,$$

because $\text{supp}(g) \cap E_n \in \mathcal{P}$. Thus, $\mathcal{E} \subset E_\rho$ and consequently

$$\text{cl}_{\|\cdot\|_\rho}(\mathcal{E}) \subset E_\rho,$$

since E_ρ is closed in L_ρ. Now, the inverse inclusion will be proved. Let $f \in E_\rho$ and $\epsilon > 0$ be given. Since $f \in E_\rho$ and X is a countable union of sets from $\mathcal{P}$, it follows that there exists a set $E \in \mathcal{P}$ such that

$$\|f1_{E'}\|_\rho < \tfrac{\epsilon}{2}, \ \text{ where } E' = X \setminus E.$$

Let us choose a sequence (g_n) of $\mathcal{P}$-simple functions such that

$$\text{supp}(g_n) \subset E \ \text{ for every } n \in \mathbb{N},$$

$$|g_n(x)| \le |f(x)| \ \text{ for all } x \in E,$$

$$g_n(x) \to f(x) \ \text{ for every } x \in E.$$

By the Lebesgue Dominated Convergence Theorem we get $\|g_n - f1_E\|_\rho \to 0$. Thus,

$$\|g_{n_0} - f1_E\|_\rho < \tfrac{\epsilon}{2} \ \text{ for a certain } n_0 \in \mathbb{N},$$

and

$$\|g_{n_0} - f\|_\rho \le \|g_{n_0} - f1_E\|_\rho + \|g_{n_0}1_{E'}\|_\rho + \|f1_{E'}\|_\rho \le \epsilon,$$

that is,

$$E_\rho \subset \text{cl}_{\|\cdot\|_\rho}(\mathcal{E}).$$

2.5 COMPACTNESS AND SEPARABILITY

We can establish now a characterization of compact subsets of the space E_ρ.

2.5.1 THEOREM. A set $D \subset E_\rho$ is conditionally compact if and only if the following conditions are satisfied:

(i)   for every $\alpha > 0$, the set function $\sup\{\rho(\alpha f, \cdot) \; ; \; f \in D\}$ is order continuous;

(ii)   for every sequence (f_n) of elements from D there exists a subsequence (f_{n_k}) and a function $f \in E_\rho$ such that $f_{n_k} \to f \; (\rho)$.

PROOF. Sufficiency. Let $f_n \in D$ for all $n \in \mathbb{N}$. By (ii) we can choose a subsequence (f_{n_k}) and a function $f \in E_\rho$ such that $f_{n_k} \to f \; (\rho)$. By (i) then $\rho\left(\alpha f_{n_k}, \cdot\right)$ are order equicontinuous for every $\alpha > 0$. From Corollary 2.4.6 it follows that

$$\|f_{n_k} - f\|_\rho \to 0.$$

Necessity. To prove (i) let us fix an $\epsilon > 0$ and sets $E_k \downarrow \emptyset$. By the conditional compactness of D we may find a finite set

$\{f_1, \dots , f_n\}$ of elements from E_ρ such that to every $f \in D$ there corresponds an index $i \in \{1, \dots , n\}$ for which

$$\|f - f_i\|_\rho < \tfrac{\epsilon}{2}.$$

Since all functions (f_i) belong to E_ρ, then it follows that there exists a natural number k_0 such that $\|f_i 1_{E_k}\|_\rho < \tfrac{\epsilon}{2}$ for $k \geq k_0$, $1 \leq i \leq n$. Hence,

$$\|f 1_{E_k}\|_\rho < \|f - f_i\|_\rho + \|f_i 1_{E_k}\|_\rho < \epsilon \text{ for } k \geq k_0.$$

This completes the proof of (i). In order to prove (ii) let us choose arbitrary $(f_n) \subset D$. Since D is conditionally compact, it follows that there exists a subsequence (f_{n_k}) of (f_n) and a function $f \in E_\rho$ such that $\|f_{n_k} - f\|_\rho \to 0$. Thus, $f_{n_k} \to f \, (\rho)$ in view of Proposition 2.3.3.

We want to pass now to the interesting problem of separability of E_ρ. We will go back to the question of separability in Section 3 of Chapter 3. It will turn out that under some additional assumptions L_ρ cannot be separable while it is essentially larger than E_ρ.

2.5.2 DEFINITION. Let Z be a subset of $M(X,S)$; the function semimodular ρ is said to be separable on Z if and only if $\|f 1_{(\cdot)}\|_\rho$ is a separable set function for each $f \in Z$, which means that there exists a countable $\mathcal{A} \subset \mathcal{P}$ such that to every $A \in \mathcal{P}$ there coresponds a sequence (A_k) of elements of $\mathcal{A}$ such that

$$\rho(\alpha f, A \triangle A_k) \to 0 \quad \text{for all } \alpha > 0.$$

2.5.3 PROPOSITION. If a linear space $L \subset L_\rho$ is separable and $\mathcal{P}$-solid (i.e. $f1_A \in L$ whenever $f \in L$ and $A \in \mathcal{P}$) then ρ is separable on L.

PROOF. Suppose to the contrary that ρ is not separable on L. There exists then a function $f \in L$, a number $\epsilon > 0$ and an uncountable family of sets $\mathcal{B} \subset \mathcal{P}$ such that $\|f1_{B \triangle B'}\|_\rho > \epsilon$ for arbitrary $B, B' \in \mathcal{B}$ such that $B \neq B'$. Thus

$$\epsilon < \|f1_{B \triangle B'}\|_\rho = \|f1_B - f1_{B'}\|_\rho.$$

Put $W = \{f1_B \; ; B \in \mathcal{B}\}$, observe then that W is an uncountable subset of L and $\|g - h\|_\rho > \epsilon$ for $g, h \in W, g \neq h$. This fact contradicts the separability of L.

Now we are able to state the main separability result for modular function spaces.

2.5.4 THEOREM. The following conditions are equivalent:

(i) E_ρ is separable,

(ii)   S is separable and ρ is separable on $\mathcal{S}$.

PROOF. (i) $\Rightarrow$ (ii)   Let us fix a set $A \in \mathcal{P}$ which is not ρ-null. Denote $W = \{r1_A ; r \in S\} \subset E_\rho$ and observe that W is separable. Define a one - to - one mapping $\phi : W \rightarrow S$ by the formula $\phi(r1_A) = r$. We conclude that ϕ is continuous; indeed,

$$\|r_m 1_A - r1_A\|_\rho \rightarrow 0$$

implies

$$r_m 1_A \rightarrow r1_A \ (\rho),$$

i.e.

$$\rho\Big(\alpha, \{x \in X; |r_n - r| \geq \epsilon\} \cap A\Big) \rightarrow 0$$

holds for every positive ϵ and α. Thus, $|r_n - r| < \epsilon$ for n sufficiently large. Finally, S is a continuous image of the separable set W and, therefore, S is separable itself. Since $\mathcal{S}$ is a $\mathcal{P}$-solid subspace of L_ρ, it follows from the previous proposition that ρ is separable on $\mathcal{S}$.

(ii) $\Rightarrow$ (i)   Let us denote by Q a countable dense subset of S and by $\mathcal{B}_{n,r}$ a countable subfamily of sets dense in $\{X_n \cap A ; A \in \mathcal{P}\}$ with respect to the pseudometric

$$(A,B) \mapsto \|r1_{X_n \cap (A \triangle B)}\|_\rho, \text{ A and B from } \mathcal{P}.$$

Let us put

$$\mathfrak{B} = \bigcup_{n \in N, r \in Q} \mathfrak{B}_{n,r}.$$

It is enough to prove that $\{r1_B \; ; \; B \in \mathfrak{B}\}$ is dense in $\{s1_A \; ; \; s \in S, A \in \mathcal{P}\}$. Fix $s \in S, A \in \mathcal{P}$ and $\epsilon > 0$. By the density of Q in S and by Proposition 2.1.5 we can choose an $r \in Q$ such that

$$\|s1_A - r1_A\|_\rho < \tfrac{\epsilon}{3}.$$

Since $A, X_n \in \mathcal{P}$ and $X_n \uparrow X$ it follows that $Y_n = A \setminus X_n \downarrow \emptyset$ and then

$$\|r1_{Y_n}\|_\rho < \tfrac{\epsilon}{3} \text{ for n sufficiently large.}$$

It follows from the assumptions that there exists a set $B \in \mathfrak{B}_{n,r}$ such that

$$\|r1_{A \cap X_n} - r1_B\|_\rho = \|r1_{X_n \cap (A \triangle B)}\|_\rho < \tfrac{\epsilon}{3}.$$

Finally,

$$\|s1_A - r1_B\|_\rho$$

$$\leq \|s1_A - r1_A\|_\rho + \|r1_{A \cap X_n} - r1_B\|_\rho + \|r1_{Y_n}\|_\rho < \epsilon.$$

The proof is fully completed.

Bibliographical remarks

The material presented here is based on papers [4] *and* [5] *by Kozlowski. As regards the historical background of the theory of modular function spaces, the reader is referred to Chapter 1. The interaction between the general theory of modular function spaces and some special cases will be discussed in Chapter 4. The bibliographical remarks after that chapter will give the necessary references.*

3. Function Semimodulars: Further Results

The results of the last two sections of the previous chapter show how important a role is played by the subspace E_ρ. Generally speaking, this space consists of all functions with order continuous F-norms. This fact suggests some new questions:

(1) When are spaces L_ρ and E_ρ equal?
(2) How can one characterize this equality in terms of order continuity?
(3) Does problem (1) have anything to do with some estimates for function semimodulars?
(4) What is the relation between problem (1) and the question of equivalence between the F-norm convergence and the modular convergence?

In this chapter we try to give some reasonable answers to the questions mentioned above.

3.1 Δ_2 - CONDITION AND EQUIVALENCE PROPERTY

3.1.1. DEFINITION. By L_ρ^0 we shall mean the class of all functions $f \in M(X,S)$ such that $\rho(f,\cdot)$ is order continuous. It is easy to prove that $L_\rho^0 \subset L_\rho$ (cf. the proof of Theorem 2.4.2). The smallest linear subspace of L_ρ which contains L_ρ^0 will be denoted by L_ρ^c, i.e. $f \in L_\rho^c$ if and only if there exists a number $\lambda > 0$ such that $\lambda f \in L_\rho^0$. Let us define L_ρ^f as the class of all functions $f \in L_\rho$ such that $\rho(f) < \infty$.

The following remark is an immediate consequence of Def. 3.1.1.

3.1.2 PROPOSITION. $E_\rho \subset L_\rho^0 \subset L_\rho^c \subset L_\rho$ and L_ρ^0 is a linear space if and only if $E_\rho = L_\rho^0 = L_\rho^c$.

Modifying slightly the proof of the Vitali Theorem we obtain the following lemma.

3.1.3 LEMMA. Let $f_n \in L_\rho^0$ and $f_n \to 0$ ρ-a.e., then the following conditions are equivalent:

(i) $\rho(f_n) \to 0$;

(ii) $\rho(f_n,\cdot)$ are order equicontinuous.

It is easy to check that L_ρ^0 is a convex and balanced subset of L_ρ. If we assume additionally that L_ρ^0 is absorbing in L_ρ then clearly $L_\rho^c = L_\rho$. In fact, in many special cases these spaces are identical. Let us note, for instance, that L_ρ^0 is absorbing in L_ρ when $L_\rho^f \subset L_\rho^0$. Throughout this chapter we will always understand that L_ρ^0 is absorbing (sometimes we will assume that $L_\rho^f \subset L_\rho^0$).

3.1.4 DEFINITION. We say that ρ satisfies the Δ_2-condition whenever the following implication is true:

(3.1.5)   if $f_n \in L_\rho^c$ for all n and $\rho(f_n, \cdot)$ are order equicontinuous then $\rho(2f_n, \cdot)$ are order equicontinuous.

3.1.6 THEOREM. The following conditions are equivalent:

(a) ρ satisfies the Δ_2-condition;

(b) L_ρ^0 is a linear subspace of L_ρ;

(c) $L_\rho = E_\rho = L_\rho^0$;

(d)   the modular convergence is equivalent to the F-norm convergence in L_ρ.

PROOF. (a) $\Rightarrow$ (b) Let $f \in L_\rho^0$ and $E_k \downarrow \emptyset$, $E_k \in \Sigma$. Since $f \in L_\rho^0$, it follows that $\rho(f, E_k) \to 0$. By Δ_2 we get $\rho(2f, E_k) \to 0$. Hence, $2f \in L_\rho^0$ and consequently L_ρ^0 is linear.

(b) $\Rightarrow$ (c) We have $E_\rho \subset L_\rho^0 \subset L_\rho$, and L_ρ is the smallest linear space containing L_ρ^0. E_ρ is the largest linear space contained in L_ρ^0. Since L_ρ^0 is linear, then $L_\rho = E_\rho = L_\rho^0$.

(c) $\Rightarrow$ (a) Suppose to the contrary that there exists a sequence of functions (f_n) from L_ρ such that $\rho(f_n, \cdot)$ are order equicontinuous while $\rho(2f_n, \cdot)$ are not. It follows then that $\rho(2f_n, \cdot)$ are not uniformly exhaustive. Thus, passing if necessary to a suitable subsequence, we may assume that there exists a sequence of disjoint sets $D_k \in \Sigma$ and a constant $\eta > 0$ such that

$$\rho(2f_k, D_k) > \eta \ \text{ for every } k \in \mathbb{N}.$$

Observe that $\rho(f_k, D_k) \to 0$ because

$$\rho(f_k, D_k) \le \sup_{n \in \mathbb{N}} \rho(f_n, D_k) \to 0.$$

The latter convergence is a consequence of the fact that $\sup_{n \in \mathbb{N}} \rho(f_n, \cdot)$ is order continuous and thus exhaustive.

Let (f_{k_i}) be a subsequence of (f_k) such that

$$\sum_{i=1}^{\infty} \rho\left(f_{k_i}, D_{k_i}\right) \le 1.$$

Denoting

$$s_m = \sum_{i=1}^{m} f_{k_i} 1_{D_{k_i}} \ \text{ and } \ f = \sum_{i=1}^{\infty} f_{k_i} 1_{D_{k_i}},$$

we get $s_m \in L_\rho = E_\rho$ and

$$\rho(f - s_m) \le \sum_{i=m+1}^{\infty} \rho\left(f_{k_i}, D_{k_i}\right) \to 0.$$

The function f is a member of $L_\rho = E_\rho$. Indeed, let $\epsilon > 0$ be given arbitrarily. Moreover, let $0 < \epsilon_k < \frac{1}{2}$, $\epsilon_k \to 0$, $\lambda_k = 2\epsilon_k < 1$. We have then

$$\rho(\epsilon_k f) \le \rho\left(\lambda_k(f - s_m)\right) + \rho\left(\lambda_k s_m\right) \le \rho\left(f - s_m\right) + \rho\left(\lambda_k s_m\right).$$

We may take m_0 such that

$$\rho\left(f - s_{m_0}\right) < \frac{\epsilon}{2}$$

and k_0 such that

$$\rho\left(\lambda_k s_{m_0}\right) < \frac{\epsilon}{2} \quad \text{for all natural } k \ge k_0.$$

Thus,

$$\rho(\epsilon_k f) < \epsilon \text{ for } k \ge k_0,$$

and then $f \in L_\rho = E_\rho$. Since $L_\rho = E_\rho$ is linear, it follows that 2f is a member of E_ρ as well. However, it was observed above, that $\rho(2f, D_k) = \rho(2f_k, D_k) > \eta$. Hence, $\rho(2f, \cdot)$ is a σ-subadditive measure which is not exhaustive, i.e. $\rho(2f, \cdot)$ must not be order continuous, which means that the function 2f does not belong to E_ρ. This contradiction completes this part of the proof.

(a) $\Rightarrow$ (d)   It suffices to prove that $\rho(f_n) \to 0$ implies

$\rho(2f_n) \to 0$ for $f_n \in L_\rho$. Assume, therefore, that $f_n \in L_\rho$ and $\rho(f_n) \to 0$. Let us take an arbitrary subsequence (h_n) of (f_n). We can choose a subsequence (g_n) of (h_n) such that $g_n \to 0$ ρ-a.e. As we proved, it follows from (a) that $L_\rho = E_\rho = L_\rho^0$. By Lemma 3.1.3 we conclude then that $\rho(g_n, \cdot)$ are order equicontinuous. Hence, $\rho(2g_n, \cdot)$ are also equicontinuous by the Δ_2-condition. Using Lemma 3.1.3 again we get $\rho(2g_n) \to 0$. By arbitrariness of (h_n) we have therefore $\rho(2f_n) \to 0$.

(d) $\Rightarrow$ (a) Let $f_n \in L_\rho$ and let $\rho(f_n, \cdot)$ be order equicontinuous. Assume to the contrary that there exists a sequence of sets $E_k \in \Sigma$ such that $E_k \downarrow \emptyset$, an $\epsilon > 0$ and a subsequence (g_n) of (f_n) such that

$$\rho(2g_k, E_k) = \rho\left(2g_k 1_{E_k}\right) > \epsilon.$$

On the other hand,

$$\rho(g_k, E_k) \leq \sup_{n \in \mathbb{N}} \rho(f_n, E_k) \to 0.$$

By (d) then

$$\rho(2g_k, E_k) = \rho\left(2g_k 1_{E_k}\right) \to 0.$$

Contradiction.

3.1.7 REMARK. In Theorem 3.1.6, we proved that equivalence between the modular convergence and the F-norm convergence may hold if and only if $E_\rho = L_\rho$. We may ask the question: what is the largest subspace of L_ρ which possesses this equivalence property? At first one might say that the space E_ρ itself should have that property. It is quite surprising that the answer to the question is negative; if ρ does not have the Δ_2-property then E_ρ does not have the equivalence property and, as it turns out, no "reasonable" subspace of L_ρ can satisfy it. First, we need some definitions.

3.1.8 DEFINITION. Let F be a linear subspace of a modular function space L_ρ. We say that F has the equivalence property if $\|f_n\|_\rho \to 0$ whenever $(f_n) \subset F$ and there is an $\alpha > 0$ such that $\rho(\alpha f_n) \to 0$.

3.1.9 DEFINITION. A closed subspace F of L_ρ will be called an ideal if and only if $g \in F$ whenever there exists a function $f \in F$ such that $|g(x)| \leq |f(x)|$ ρ-a.e. An ideal $F \subset L_\rho$ will be called super order dense if and only if to every $f \in L_\rho$ there corresponds a sequence of (f_n) of functions from F such that

$$|f_n(x)| \uparrow |f(x)| \quad \rho\text{-a.e.}$$

3.1.10 THEOREM. If ρ is a Fatou function semimodular, $F \subset L_\rho$ has the equivalence property and there exists an ideal A which is super order dense in L_ρ, and is contained in F, then ρ satisfies the Δ_2-condition.

PROOF. Assume to the contrary that ρ does not satisfy the Δ_2-condition. In view of Theorem 3.1.6 there exist then a positive number ϵ and a sequence of functions $f_n \in L_\rho$ such that $\rho(f_n) \to 0$ while $\|f_n\|_\rho > \epsilon$. Since A is super order dense in L_ρ it follows that to every $n \in N$ there exists a sequence $(s_{n,k})$ of functions from A such that $|s_{n,k}(x)| \uparrow |f_n(x)|$ ρ-a.e., for every $n \in N$. By the Fatou property, to every $n \in N$ there corresponds a natural number k_n such that

$$\|f_n\|_\rho \leq \|s_{n,k}\|_\rho + \tfrac{\epsilon}{2} \text{ for } k \geq k_n.$$

Denoting $g_n = s_{n,k_n}$ we get

$$0 \leq \rho(g_n) \leq \rho(f_n) \to 0.$$

Since $g_n \in A \subset F$, then $\|g_n\|_\rho \to 0$. Hence,

$$\epsilon < \|f_n\|_\rho \leq \|g_n\|_\rho + \tfrac{\epsilon}{2} \to \tfrac{\epsilon}{2} \text{ as } n \to \infty.$$

Contradiction.

Let us observe that in place of A one may take the smallest ideal containing the space of all $\mathcal{P}$-simple functions $\mathcal{E}$, the ideal of all measurable bounded functions with supports from $\mathcal{P}$. In this context, Theorem 3.1.10 can be reformulated as follows:

3.1.11 PROPOSITION. If ρ is a Fatou semimodular which does not satisfy the Δ_2-condition, then no ideal which contains $\mathcal{E}$ may possess the equivalence property.

3.2 ATOMLESS FUNCTION SEMIMODULARS

Before we consider other problems connectected with the Δ_2-property, we have to introduce some new notions which will also be of some use in dealing with the question of existence of nontrivial linear functionals over modular function spaces.

3.2.1 DEFINITION. A set $A \in \Sigma$ is called a ρ-atom if and only if every $B \in \Sigma$, $B \subset A$ is either ρ-null or there holds $\rho(f,B) = \rho(f,A)$ for all $f \in \mathcal{E}$. We say that ρ is atomless if there are no ρ-atoms in the σ-algebra Σ.

3.2.2 REMARK. If ρ is disjointly additive and atomless then to every f $\in$ $\mathcal{S}$ there exist two disjoint sets X_1, $X_2 \in \Sigma$ such that

$$X_1 \bigcup X_2 = X \text{ and } \rho(f,X_1) = \rho(f,X_2) = \tfrac{1}{2}\,\rho(f).$$

PROOF. $\rho(f,\cdot)$ is an atomless measure because it is an order continuous and disjointly additive set function. The rest of the proof follows in a classical way.

The above remark justifies the following definitions.

3.2.3 DEFINITION. We say that a function semimodular ρ is of finite type if and only if there is a universal constant $k \in \mathbb{N}$, $k \geq 2$, such that to every function f $\in$ $\mathcal{S}$ there corresponds $\{\, X_1(f), \ldots , X_k(f) \,\}$, a partition of X (depending on f), with

$$\rho\Big(f,X_i(f)\Big) \leq \frac{\rho(f)}{2}$$

for i = 1, ... , k. The smallest number $k \in \mathbb{N}$ with this property will be called an index of ρ and will be denoted by K_ρ.

3.2.4 PROPOSITION. If ρ is a function semimodular of finite type then ρ is atomless.

PROOF. Assume to the contrary that there is a ρ-atom A in Σ. Put $f = r1_A$, where $r \in S$ is arbitrary. Since ρ is of finite type, there exists a partition $\{ X_1(f), \dots , X_k(f) \}$ such that

$$\rho\left(f, X_i(f)\right) \le \frac{\rho(f)}{2}.$$

Put $A_i = A \cap X_i(f)$. Then,

$$0 < \rho(f) \le \sum_{i=1}^{k} \rho(f,A_i).$$

Hence, $\rho\left(f,A_{i_0}\right) > 0$ for an index $i_0 \in \{1, \dots , k\}$. On the other hand,

$$\rho\left(f, A_{i_0}\right) \le \frac{\rho(f)}{2} < \rho(f).$$

We conclude that A cannot be a ρ-atom. Contradiction.

It is clear that if ρ is of finite type then to every $f \in \mathcal{E}$ there corresponds a sequence of sets $Y_i \in \Sigma$ such that

$$\rho(f,Y_i) \le \frac{\rho(f)}{4} \quad \text{for } i = 1, \dots , (K_\rho)^2.$$

This procedure can certainly be continued.

3.2.5 EXAMPLE. Let $(X_p)_{p\in\mathbb{N}}$ be a disjoint partition of the interval $[0,1]$ and let $m(X_p) = 2^{-p}$, where m denotes the Lebesgue measure in $[0,1]$. Let $\mathcal{P}_r$ $(r\in\mathbb{N})$ be a δ-ring generated by the sets of the form $A \cap X_p$ for $p \in \{1, \dots , r\}$, $A \subset [0,1]$ being measurable. For a measurable function f,

$$\rho_r(f,E) = \sum_{p=1}^{r} \left(\int_{X_p \cap E} |f|^P \, dm \right)^{\frac{1}{p}} + \max_{p=1,\,\ldots,\,r} \left\{ \int_{X_p \cap E} |f|^P \, dm; \right\}.$$

It is easy to see that ρ_r is a function modular. Put $u_p = 2 \cdot 1_{X_p}$ for $p = 1, \ldots, r$. We have

$$\rho_r(u_p) = \left(\int_{X_p} 2^P \, dm \right)^{\frac{1}{p}} + \int_{X_p} 2^P \, dm = 2.$$

We will show that for every $r \geq 3$ the index of ρ_r is greater than 2. Let us denote the index K_ρ by k and let $n = k^2$. It is easy to prove that sums and maximums of semimodulars of finite type are of finite type again. Assume now that there exists a partition $Z_1, \ldots, Z_n$ of X_r such that

$$\rho_r(u_r, Z_i) \leq \tfrac{2}{4} = \tfrac{1}{2} \quad \text{for } i = 1, \ldots, n.$$

In particular

$$\left(\int_{Z_i} 2^r \, dm \right)^{\frac{1}{r}} \leq \tfrac{1}{2},$$

i.e.

$$2 \, [m(Z_i)]^{\frac{1}{r}} \leq \tfrac{1}{2}.$$

Hence,

$$m(Z_i) \leq \left(\tfrac{1}{2} \right)^r \cdot \left(\tfrac{1}{2} \right)^r = m(X_r) \cdot \left(\tfrac{1}{2} \right)^r.$$

But

$$m(X_r) = \sum_{i=1}^{n} m(Z_i) \leq n \cdot m(X_r) \cdot \left(\tfrac{1}{2} \right)^r.$$

Thus $k^2 = n \geq 2^r$, which yields $k \geq 2^{\frac{r}{2}} > 2$ for $r \geq 3$.

This example shows that it may happen that the index of ρ is greater than 2. This property of function semimodulars will play an interesting role in our discussion of linear functionals. We will slightly modify the previous situation in order to obtain an example of an atomless semimodular which is not of finite type.

3.2.6 EXAMPLE. Let $X = [0,1]$, $(X_p)_{p \in \mathbb{N}}$ be a countable disjoint partition of X such that $m(X_p) = 2^{-p}$. Let $\mathcal{P}$ be a δ-ring generated by the sets of the form $A \cap X_p$ for all $p \in \mathbb{N}$ and all measurable sets $A \subset [0,1]$. For a measurable function $f : X \to \mathbb{R}$ we put

$$\rho(f,E) = \sum_{p=1}^{\infty} \left(\int_{X_p \cap E} |f|^p \, dm \right)^{\frac{1}{p}} + \sup \left\{ \int_{X_p \cap E} |f|^p \, dm \; ; \; p \in \mathbb{N} \right\}.$$

We can verify that ρ is a function modular. Furthermore, for $f \in L_\rho$ there holds

$$\|f\|_\Sigma = \sum_{p=1}^{\infty} \left(\int_{X_p} |f|^p \, dm \right)^{\frac{1}{p}} < \infty.$$

Indeed, if $f \in L_\rho$ then $\rho(\lambda f) \to 0$ as $\lambda \to 0$. Since $\|\cdot\|_\Sigma$ is positively homogeneous it follows that $\|f\|_\Sigma < \infty$, which implies on its own that

$$\left(\int_{X_p} |f|^P \, dm \right)^{\frac{1}{p}} < 1 \ \text{ for p sufficiently large.}$$

As the result of this inequality we get

$$\int_{X_p} |f|^P \, dm \ \leq \left(\int_{X_p} |f|^P \, dm \right)^{\frac{1}{p}} \to 0.$$

Hence,

$$\sup \left\{ \int_{X_p} |f|^P \, dm \ ; \ p \in \mathbb{N} \right\} = \int_{X_r} |f|^r \, dm$$

for a certain $r \in \mathbb{N}$. We can repeat now the same argument as was used in the previous example replacing ρ_r by ρ . We note that

$$K_\rho \geq 2^{\frac{r}{2}} \ \text{for arbitrary } \ r \in \mathbb{N}.$$

Consequently, the index K_ρ must not be finite, i.e. ρ is not of finite type.

3.3 Δ_2'- CONDITION

Let us go back to the problems connected with the Δ_2-condition. Our concept of this condition, though useful and structural, cannot be applied in many situations when some numerical estimates are wanted. This is why we shall need. another definition.

3.3.1 DEFINITION. A function semimodular ρ is said to satisfy the Δ_2'-condition if and only if to every $d > 0$ there corresponds a positive number $c(d)$ such that

$$\rho(f + g) \leq c(d) \quad \text{whenever} \quad \max\left\{\rho(f), \rho(g)\right\} \leq d.$$

Let us put

$$B(d) = \{f \in L_\rho \; ; \; \rho(f) \leq d \} \text{ and } B^1(d) = B(d) \cap \mathcal{E}.$$

The Δ_2'-condition will be used later in this section in considering the relation between separability and order continuity in modular function spaces (cf. Theorems 3.3.10 and 3.3.13), and in Chapter 7 that notion will be used to describe some approximation properties of function modulars (see Theorem 7.3.14). It is clear that Δ_2' implies always Δ_2; the first step in understanding the inverse problem was proved by Musielak (see Musielak [1], Th. 6.2). Namely, the following is true:

3.3.2 THEOREM. If ρ is a Δ_2-semimodular then there exists a $d > 0$ such that

$$\sup\{\rho(f + g) \; ; \; f, g \in B(d)\} = c(d) < \infty.$$

This partial result, however, does not answer the question of when both conditions are equivalent. First, we will use one of our examples to show that this equivalence does not hold in general.

3.3.3 PROPOSITION. The modular ρ defined in Example 3.2.6 satisfies Δ_2 but does not satisfy the Δ_2'-condition.

PROOF. Let us prove first that ρ does not satisfy Δ_2'. Indeed, let us put again $u_p = 2 \cdot 1_{X_p}$, $p \in \mathbb{N}$. We already checked that $\rho(u_p) = 2$. On the other hand

$$\rho(2u_p) \geq \int_{X_p} 2^p \cdot 2^p \, dm = 2^p \to \infty \quad \text{as} \;\; p \to \infty.$$

In order to get Δ_2 it suffices to prove that $L_\rho^0 = L_\rho$. Let $f \in L_\rho$, $E_k \downarrow \emptyset$. Since $f \in L_\rho$, $\rho(\lambda f) \to 0$ as $\lambda \to 0$, that implies

$$\sum_{p=1}^{\infty} \left(\int_{X_p} |f|^p \, dm \right)^{\frac{1}{p}} < \infty.$$

From the above, we are going to obtain that

$$\int_{X_p} |f|^p \, dm \;\to\; 0 \;\text{ as } p \to \infty.$$

Let us fix a positive ϵ and pick an index p_0 such that

$$\int_{X_p} |f|^p \, dm \; < \epsilon \text{ for } p > p_0.$$

For every $k \in \mathbb{N}$ and $1 \leq p \leq p_0$,

$$\int_{X_p \cap E_k} |f|^p \, dm \; \leq \max\left\{ \int_{X_r \cap E_k} |f|^r \, dm \; ; r = 1, \dots, p_0 \right\}.$$

There exists then $k_0 \in \mathbb{N}$ such that

$$\max\left\{ \int_{X_r \cap E_k} |f|^r \, dm \; ; r = 1, \dots, p_0 \right\} < \epsilon \text{ for } k \geq k_0.$$

Finally,

(i) $$\lim_{k \to \infty} \sup\left\{ \int_{X_p \cap E_k} |f|^p \, dm \; ; p \in \mathbb{N} \right\} = 0.$$

Since

$$\sum_{p=1}^{\infty} \left(\int_{X_p} |f|^p \, dm \right)^{\frac{1}{p}} < \infty.$$

it follows that

(ii) $$\lim_{k \to \infty} \sum_{p=1}^{\infty} \left(\int_{X_p \cap E_k} |f|^p \, dm \right)^{\frac{1}{p}}$$

$$= \sum_{p=1}^{\infty} \lim_{k \to \infty} \left(\int_{X_{p \cap E_k}} |f|^P \, dm \right)^{\frac{1}{p}} = 0.$$

By (i) and (ii) then $\rho(f, E_k) \to 0$ as $k \to \infty$, i.e. $f \in L_\rho^0$, which means that ρ satisfies the Δ_2-condition.

The previous proposition, as well as the next "positive" result, shows the importance of being of finite type for the equivalence between Δ_2 and Δ_2'.

3.3.4 LEMMA. Assume that ρ is of finite type and that $L_\rho^f \subset L_\rho^0$. If ρ does not satisfiy Δ_2' then there exists a function $g \in L_\rho^0$ such that $\beta f \notin L_\rho^0$ for every $\beta > 1$.

PROOF. For the sake of clarity we divide the proof into several steps.

Step I. The following implies Δ_2':

(3.3.5)   for every $d > 0$ there holds
$$M_d = \sup\{\rho(2f) \; ; \; f \in B(d)\} < \infty.$$

Indeed, assuming (3.3.5) we have

$$\rho(f + g) \leq \rho(2f) + \rho(2g) \leq 2M_d = c(d) < \infty.$$

Step II. To get (3.3.5) it is enough to obtain

(3.3.6)   for every d > 0 there holds
$$\sup\{\rho(2f) ; f \in B^1(d)\} < \infty.$$

Let $f \in B(d)$. Since ρ satisfies the Δ_2-condition, it follows that

$$L_\rho = E_\rho = cl_{\|\cdot\|_\rho}(\mathcal{E})$$

and, therefore, there exists a sequence (f_n) of functions $f_n \in \mathcal{E}$ such that

$$\|f_n - f\|_\rho \to 0.$$

Then

$$\rho\left(4(f_n - f)\right) \to 0 \text{ as } n \to \infty.$$

Hence, for n sufficiently large

$$\rho(2f) \leq \rho\left(4(f_n - f)\right) + \rho(4f_n) \leq 1 + \rho(4f_n)$$

and

$$\rho\left(\tfrac{1}{2} f_n\right) \leq \rho(f_n - f) + \rho(f) \leq 1 + d.$$

By (3.3.6) we have

$$\rho(f_n) = d_1 = c(1 + d)$$

and consequently

$$\rho(2f_n) \leq d_2 = c(d_1) \text{ and } \rho(4f_n) \leq c(d_2).$$

Thus,

$$\rho(2f) \leq 1 + c(d_2) < \infty.$$

Putting

$$M = 1 + c(d_2),$$

we get

$$\sup\{\rho(2f) ; f \in B(d)\} \leq M < \infty.$$

Step III. Let us observe that (3.3.6) is implied by the following condition:

(3.3.7)   To every $d > 0$ there exists a constant $0 < M < \infty$ such that for every $n \in \mathbb{N}$

$$\sup\{\rho(\lambda_n f) ; f \in B^1(d) \} \leq M, \text{ where } \lambda_n = 1 + \tfrac{1}{n}.$$

Step IV. Suppose that Δ_2' does not hold; there exists then a number $d > 0$ such that to every natural n there corresponds a function $f_n \in B^1(d)$ and a number $\gamma_n = \lambda_{k_n}$ for which $\rho(\gamma_n f_n) > 2^n$. Let $n_1 > 3$ be such that

(3.3.8) $2^{n_1-3} \geq k^2$ where $k = K_\rho.$

Let us take a partition $B_1^{(1)}, \dots, B_k^{(1)}$ such that

(3.3.9)
$$\rho\left(f_{n_1}, B_i^{(1)}\right) \le \frac{\rho(f_{n_1})}{2} \le \frac{d}{2}.$$

Thus,

$$\rho\left(\gamma_{n_1}f_{n_1}, B_{i_0}^{(1)}\right) \ge 1$$

for a certain $i_0 \in \{1, 2, \ldots, k^2\}$. Indeed, if

$$\rho\left(\gamma_{n_1}f_{n_1}, B_i^{(1)}\right) < 1$$

held for every $i \in \{1, 2, \ldots, k^2\}$ then we would get

$$2^{n_1} < \rho\left(\gamma_{n_1}f_{n_1}\right) \le \sum_{i=1}^{k^2} \rho\left(\gamma_{n_1}f_{n_1}, B_i^{(1)}\right) < k^2 \le 2^{n_1 - 3},$$

which is impossible. Since $\gamma_{n_1}f_{n_1} \in \mathcal{E}$, $\rho\left(\gamma_{n_1}f_{n_1}, B_{i_0}^{(1)}\right)$ and ρ is, in view of Prop. 3.2.4, atomless, we may choose a measurable set

$$D_1 \subset B_{i_0}^{(1)}$$

such that

(3.3.10)
$$1 \le \rho\left(\gamma_{n_1}f_{n_1}, D_1\right) \le 2^{n_1 - 2}.$$

Observe that one of the following conditions must be satisfied:

(3.3.11)
$$\rho\left(\gamma_{n_1}f_{n_1+m}, D_1\right) \ge 2^{n_1 + m - 1}$$

for almost all natural m, or

(3.3.12) $\rho\left(2f_{n_1+m}, X \setminus D_1\right) \geq 2^{n_1+m-1}$

for almost all natural numbers m. If both (3.3.11) and (3.3.12) do not hold, then

$$2^{n_1+m} < \rho\left(\gamma_{n_1} f_{n_1+m}\right) \leq 2^{n_1+m}$$

for m sufficiently large. Contradiction. In case (3.3.12) let us put $A_1 = D_1$, $g_1 = f_{n_1}$. In case (3.3.11) the following inequality holds:

(3.3.13) $2^{n_1-2} \leq \rho\left(\gamma_{n_1} f_{n_1}, X \setminus D_1\right).$

Indeed, if

$$\rho\left(\gamma_{n_1} f_{n_1}, X \setminus D_1\right) < 2^{n_1-2}$$

then by (3.3.10)

$$2^{n_1} < \rho\left(\gamma_{n_1} f_{n_1}\right) \leq \rho\left(\gamma_{n_1} f_{n_1}, D_1\right) + \rho\left(\gamma_{n_1} f_{n_1}, X \setminus D_1\right)$$

$$< 2^{n_1-2} + 2^{n_1-2} = 2^{n_1-1},$$

which is not possible. Let us denote $D' = X \setminus D$. By the finite type of ρ, similarly as was shown previously, to the function $f_{n_1} 1_{D'}$ there exists a set

$$A_1 \in \Sigma, A_1 \subset D'$$

such that

$$\rho\left(\gamma_{n_1} f_{n_1}, A_1\right) \geq 1$$

and

$$\rho\left(f_{n_1}, A_1\right) \leq \frac{d}{2}.$$

Put $g_1 = f_{n_1}$.

Step V. Set $Y_1 = X \setminus A_1$. We observed that

$$\rho\left(\gamma_{n_1} f_{n_1+m}, Y_1\right) \geq 2^{n_1+m-1}$$

for m sufficiently large and $n_1 \geq 3$. Let $n_2 \geq n_1$ be such a natural number that

$$2^{n_2-3} \geq k^3.$$

Let us choose a natural number m_2 such that

$$\rho\left(\gamma_{n_1} f_{n_1+m_2}, Y_1\right) > 2^{n_2}$$

and

$$2^{n_2-3} \geq k^3.$$

Now we can repeat the procedure from Step IV. By a simple induction we obtain the sequences $g_n \in B^1(d)$, (β_n) a subsequence of (γ_n) and $A_n \in \Sigma$ (mutually disjoint) for which

$$\rho\left(\beta_n g_n, A_n\right) \geq 1 \text{ and } \rho\left(g_n, A_n\right) \leq \frac{d}{2^n}.$$

Define

$$g = \sum_{n=1}^{\infty} g_n 1_{A_n}.$$

Since for every $\beta > 1$ there exists $n_0 \in \mathbb{N}$ such that $\beta \geq \beta_n$ for $n \geq n_0$, we have

$$\rho(\beta g, A_n) \geq \rho(\beta_n g, A_n) = \rho(\beta_n g_n, A_n) \geq 1.$$

It follows that $\rho(\beta g, \cdot)$ is not an exhaustive subadditive measure, i.e. it is not order continuous. Hence, $2g \notin L_\rho^0$. On the other hand

$$(3.3.14) \qquad \rho(g) \leq \sum_{n=1}^{\infty} \rho(g_n, A_n) \leq d \sum_{n=1}^{\infty} \frac{1}{2^n} = d,$$

which implies that $g \in L_\rho^f \subset L_\rho^0$. The proof of Lemma 3.3.4 is fully completed.

As an immediate but important conclusion, we obtain the following result which answers the question of the equivalence between Δ_2 and Δ_2'

3.3.15. THEOREM. Assume that ρ is of finite type and $L_\rho^f \subset L_\rho^0$. If ρ satisfies Δ_2 then ρ satisfies Δ_2' as well.

We shall use the Δ_2'condition and Theorem 3.3.15 in order to obtain more complete charcterizations of spaces E_ρ and L_ρ for an interesting class of function semimodulars. The following technical lemma will play an interesting role in proving Theorem 3.3.20.

3.3.16 LEMMA. Let ρ be an s-convex function semimodular of finite type and let $L_\rho^f \subset L_\rho^0$. If ρ does not satisfy the Δ_2-condition then

$$(3.3.17) \quad \Pi(E_\rho) = \left\{ f \in L_\rho \; ; \; d(f,E_\rho) = \inf_{g \in E_\rho} \|f - g\|_\rho < 1 \right\} \text{ is a}$$

proper subset of L_ρ^0.

PROOF. Let $f_0 \in E_\rho$; we will prove that if $f \in L_\rho$ and

$$\|f - f_0\|_\rho < 1$$

then $f \in L_\rho^0$. Indeed, let $\|f - f_0\|_\rho < 1$; then we can find a number $\alpha \in (0,1)$ such that

$$\|f - f_0\|_\rho < (1 - \alpha)^s.$$

Hence,

$$\left\| \frac{f - f_0}{1 - \alpha} \right\|_\rho < 1$$

and therefore

$$\rho\left(\frac{f - f_0}{1 - \alpha} \right) \le \left\| \frac{f - f_0}{1 - \alpha} \right\|_\rho < 1.$$

We conclude that

$$\frac{f - f_0}{1 - \alpha} \in L_\rho^f \subset L_\rho^0.$$

Observe that $\frac{f_0}{\alpha} \in L_\rho^0$ because $f_0 \in E_\rho$. Then

$$f = (1 - \alpha) \, \frac{f - f_0}{1 - \alpha} + \alpha \, \frac{f_0}{\alpha}$$

belongs to L_ρ^0 because L_ρ^0 is a convex subset of L_ρ. It remains to prove that the inclusion in (3.3.17) is proper. By Lemma 3.3.4 there exists a function $g \in L_\rho^0$ such that $\beta g \notin L_\rho^0$ for every $\beta > 1$. We shall show that $g \notin \Pi(E_\rho)$. In fact, if $d(g, E_\rho)$ were less than unity, then we would find a $\beta > 1$ such that

$$d(\beta g, E_\rho) = \inf_{h \in E_\rho} \|\beta(g - h)\|_\rho = \beta^S \inf_{h \in E_\rho} \|\beta(g - h)\|_\rho < 1.$$

This would mean that $\beta g \in L_\rho^0$ which is not true. We have thus proved that $\Pi(E_\rho)$ is a proper subset of L_ρ^0.

3.3.18 DEFINITION. A function semimodular ρ is said to be locally determined if for every $B \subset \mathcal{S}$ such that

$$\sup\{\rho(f) \; ; \; f \in B\} = \infty$$

there exists a set $Z \in \mathcal{P}$ such that for every measurable nonnull $Y \subset Z$

$$\sup\{\rho(f,Y) \; ; \; f \in B\} = \infty.$$

To avoid some pathologies, to the end of this section we restrict our consideration to the case of locally determined function semimodulars. It is easy to show that many classical function modulars satisfy this condition. It follows by (3.3.6) that if ρ is locally determined and does not satisfy Δ_2' then there is a $Z \in \mathcal{P}$ such that ρ_Y does not satisfy Δ_2' for any measurable nonnull sets $Y \subset Z$, where $\rho_Y(f,A) = \rho(f,A \cap Y)$, for functions $f \in L_\rho$ such that $\mathrm{supp}(f) \subset Y$. The modular function space induced by ρ_Y will be denoted by $L_\rho(Y)$.

To the end of this section we assume that

(1) ρ is s-convex;

(2) ρ is of finite type;

(3) ρ is locally determined;

(4) ρ is separable on $\mathcal{E}$.

The next result shows that under the above assumptions the separability of L_ρ and the condition Δ_2 are equivalent.

3.3.20 THEOREM. Let ρ be a function semimodular described above. If S is a separable Banach space and $L_\rho^f \subset L_\rho^0$ then the following conditions are equivalent:

(3.3.21) L_ρ is separable;

(3.3.22) ρ satisfies Δ_2.

PROOF. $(3.3.22) \Rightarrow (3.3.21)$ If ρ satisfies Δ_2 then by Theorem 3.1.6 $L_\rho = E_\rho$. In view of Theorem 2.5.4, the subspace E_ρ is separable; then L_ρ is separable.

$(3.3.21) \Rightarrow (3.3.22)$ Suppose to the contrary that Δ_2 does not hold. Thus ρ does not satisfy Δ_2' But ρ is locally determined; then in view of Remark 3.3.19 we get a nonnull set $Z \in \mathcal{P}$ such that ρ_Y does not satisfy Δ_2' for every $Y \subset Z$. Let us denote by Q the countable dense subset of L_ρ and let us fix $\alpha > \alpha_0$. There exists a number $\epsilon > 0$ such that $\rho(\alpha, Z) > \epsilon$. Let us observe that to every $f_k \in Q$ there exists a set $X_k \subset Z$ such that f_k is bounded on X_k and

$$\rho(\alpha, Z \setminus X_k) < \frac{\epsilon}{2^k}.$$

Defining

$$Y = \bigcap_{k=1}^{\infty} X_k$$

we observe that every function f_k from Q is bounded on Y and

$$\rho(\alpha, Z \setminus Y) = \rho\left(\alpha, Z \setminus \bigcap_{k=1}^{\infty} X_k\right) = \rho\left(\alpha, \bigcup_{k=1}^{\infty} (Z \setminus X_k)\right)$$

$$\leq \sum_{k=1}^{\infty} \rho(\alpha, Z \setminus X_k) \leq \sum_{k=1}^{\infty} \frac{\epsilon}{2^k} = \epsilon.$$

Since

$$\epsilon < \rho(\alpha, Z) \leq \rho(\alpha, Z \setminus Y) + \rho(\alpha, Y) \leq \epsilon + \rho(\alpha, Y),$$

it follows that $\rho(\alpha, Y) \neq 0$, i.e. Y is not ρ-null because $\alpha > \alpha_0$. By the hypothesis, ρ_Y does not satisfy Δ_2'. Therefore, by Lemma 3.3.4, there exists $g_0 \in L_\rho^0(Y)$ such that $\beta g_0 \notin L_\rho^0(Y)$ for all $\beta > 1$. Let us observe that g_0 does not belong to $\Pi(E_\rho)$. Indeed, if this were the case, then $\beta_0 g_0 \in \Pi(E_\rho)$ for some $\beta_0 > 1$ and, by Lemma 3.3.16, $\beta_0 g_0 \in L_\rho^0$, which contradicts the definition of g_0. Let us put $h_0 = g_0 1_Y$ and $g_n = f_n 1_Y$ for $f_n \in Q$. Let us observe that $h_0 \in L_\rho$ and that g_n, as bounded functions with supports from $\mathcal{P}$, belong to E_ρ. Since $g_0 \notin \Pi(E_\rho)$, then

$$\|g_0 - g_n\|_\rho \geq 1 \text{ for all } g_n.$$

Then

$$\|h_0 - f_n\|_\rho \geq \|(h_0 - f_n)1_Y\|_\rho = \|g_0 - g_n\|_\rho \geq 1.$$

This means that the sequence (f_n) is not dense in L_ρ. The contradiction completes the proof.

Combining Theorem 3.3.20 with Theorem 3.1.6 we obtain more complete characterization of the role played by the subspace E_ρ and the condition Δ_2.

3.3.23 THEOREM. Under the assumption of Theorem 3.3.20 each of the following condition implies the other:

(a) L_ρ is separable;

(b) $E_\rho = L_\rho^0 = L_\rho$;

(c) L_ρ^0 is a linear subspace of L_ρ;

(d) ρ satisfies the Δ_2-condition;

(e)   the modular and the F-norm convergence are

           equivalent in L_ρ.

In particular, the assumptions of the above theorem are satisfied for Musielak-Orlicz spaces (cf. Propositions 4.1.11 and 4.1.12). The relation between Δ_2 and separability of L_ρ was known for classical Orlicz spaces (see the book of Krasnosel'skii and Rutickii [1]).

3.4 LINEAR FUNCTIONALS

At first look it may be surprising, but it turns out that the notion of index of a function semimodular is quite useful in considering the question of existence of nontrivial linear continuous functionals over modular function spaces. Let us start with the following "negative" result.

3.4.1 THEOREM. Let ρ be a Δ_2-function semimodular of finite type. Suppose that for every $f \in \mathcal{E}$ there holds

$$\frac{\rho(k^n f)}{2^n} \to 0 \text{ as } n \to \infty, \text{ where } k = K_\rho.$$

Then, there is no nontrivial linear continuous functional over the space L_ρ.

PROOF. Let Λ be a continuous linear functional over L_ρ. Denote $W = \{ g \in L_\rho \; ; \; |\Lambda g| \leq 1 \}$. Hence, W is a convex subset of L_ρ. Since Λ is continuous, then W is a neighborhood of zero but ρ satisfies the Δ_2-condition; it follows then that there exists an $\epsilon > 0$ such that $\{ g \in L_\rho \; ; \; \rho(g) \leq \epsilon \} \subset W$. Indeed, assume to the contrary that there exist $g_n \in L_\rho$ $(n \in \mathbb{N})$ such that

$$\rho(g_n) \leq \tfrac{1}{n} \text{ but } |\Lambda g_n| > 1.$$

Then, $\rho(g_n) \to 0$ and by Δ_2 we get $\|g_n\|_\rho \to 0$. Hence, in view of the continuity of Λ, we get $|\Lambda g_n| \to 0$. Contradiction. Fix an arbitrary simple function $f \in \mathcal{E}$. We may write

$$f = \frac{1}{k^n} \sum_{i=1}^{k^n} k^n f 1_{A_i},$$

where $A_1, \dots, A_{k^n} \in \mathcal{P}$ are mutually disjoint,

$$\bigcup_{i=1}^{k^n} A_i = \mathrm{supp}(f) \in \mathcal{P}$$

and

$$\rho\left(k^n f 1_{A_i} \right) \leq \frac{\rho(k^n f)}{2^n} \text{ for } i = 1, 2, \dots, k^n.$$

Since

$$\frac{\rho(k^n f)}{2^n} \to 0 \text{ as } n \to \infty,$$

then for n sufficiently large we have

$$\frac{\rho(k^n f)}{2^n} \leq \epsilon,$$

which implies

$$k^n f 1_{A_i} \in W \text{ for } i = 1, 2, \dots , k^n.$$

Since W is convex, it follows that

$$f = \frac{1}{k^n} \sum_{i=1}^{k^n} k^n f 1_{A_i} \in W,$$

i.e. $|\Lambda f| \leq 1$. By the Δ_2 and since $\mathcal{E}$ is dense in L_ρ, we get that $|\Lambda g| \leq 1$ for every $g \in L_\rho$ which means that $\Lambda = 0$.

3.4.2 REMARK. Theorem 3.4.1 gives an interesting characterization of functionals which are both function semimodulars of finite type and s-homogeneous F-norms ($s \leq 1$). Since

$$\rho(k^n f) = (k^n)^s \rho(f),$$

then the condition

$$\frac{\rho(k^n f)}{2^n} \to 0 \text{ as } n \to \infty, \text{ where } k = K_\rho,$$

is equivalent to $k < 2^{\frac{1}{s}}$. Obviously, this cannot hold if $s = 1$, i.e. if ρ is a norm. The above inequality is satisfied, however, when $k = 2$ (e.g. ρ is disjointly additive) and $s < 1$. This coincides, for instance, with well known results obtained for the case of L^s for s < 1. In the case of Musielak-Orlicz spaces, we have $K_\rho = 2$ and the condition

$$\frac{\rho(k^n f)}{2^n} \to 0$$

follows immediately from $\frac{\phi(t,u)}{u} \downarrow 0$ as u $\uparrow \infty$ (for almost all t ∈ X) and $\phi(\cdot,u)$ is integrable for all u ≥ 0.

3.4.3 THEOREM. Let ρ be an atomless, disjointly additive semimodular which satisfies the Δ_2-condition. Assume now that there exists a positive finite measure μ such that $f_n \to 0$ (μ) whenever $\rho(f_n) \to 0$ and for every D ∈ $\mathcal{P}$

(3.4.4) $$\liminf_{\alpha \to \infty} \frac{\rho(\alpha 1_D)}{\alpha \mu(D)} > 0.$$

There exists then a nontrivial linear continuous functional over L_ρ.

PROOF. It is enough to prove that L_ρ is continuously embedded into $L^1(\mu)$. Since $\mathcal{E}$ is dense in L_ρ it suffices to show that $\mathcal{E}$ is continuously embedded into $L^1(\mu)$. It follows from the assumptions that there exist $\beta > 0$ and $\alpha_0 > 0$ such that

$$\frac{\rho(\alpha 1_D)}{\alpha \mu(D)} \geq \beta \text{ for all } \alpha \geq \alpha_0 \text{ and every } D \in \mathcal{P}.$$

Let $f_n \in \mathcal{S}$ and $\|f_n\|_\rho \to 0$; then $\rho(f_n) \to 0$ and consequently $f_n \to 0$ (μ). Put $E_n = \{x \in X ; |f_n(x)| \le \alpha_0\}$ and $F_n = X \setminus E_n$. We observe that

$$\lim_{n \to \infty} \int_{E_n} |f_n| \, d\mu \to 0.$$

Since f_n is a simple function, $|f_n 1_{F_n}|$ may be written in the form

$$|f_n 1_{F_n}| = \sum_{k=1}^{m} \alpha_k 1_{A_k \cap F_n}, \text{ where } \alpha_k > \alpha_0 \ge 0 \text{ and } A_k \in \mathcal{P}.$$

Compute

$$|f_n(x) 1_{F_n \cap A_k}(x)| \le \alpha_k \le \frac{1}{\beta} \frac{\rho\left(\alpha_k 1_{F_n \cap A_k}\right)}{\mu\left(F_n \cap A_k\right)}$$

and

$$\int_{F_n \cap A_k} |f_n| \, d\mu \le \frac{1}{\beta} \rho\left(\alpha_k 1_{F_n \cap A_k}\right).$$

Thus, in view of orthogonality of ρ we have

$$\int_{F_n} |f_n| \, d\mu = \sum_{k=1}^{m} \int_{F_n \cap A_k} |f_n| \, d\mu \le \frac{1}{\beta} \sum_{k=1}^{m} \rho\left(\alpha_k 1_{F_n \cap A_k}\right)$$

$$= \frac{1}{\beta} \rho\left(f_n 1_{F_n}\right) \le \frac{1}{\beta} \rho(f_n) \to 0.$$

Hence,

$$\int_X |f_n|\ d\mu = \int_{F_n} |f_n|\ d\mu + \int_{E_n} |f_n|\ d\mu \to 0,$$

which completes the proof.

3.4.5 REMARK. For Musielak-Orlicz modulars

$$\rho(f) = \int_X \phi\big(x, |f(x)|\big)\ d\mu,$$

the implication $\rho(f_n) \to 0 \Rightarrow f_n \to 0\ (\mu)$ clearly holds, and the condition (3.4.4) is equivalent to the condition:

(3.4.6)   To every $\alpha_0 > 0$ there exists a number $\beta > 0$ such that for every $\alpha \geq \alpha_0$ and all $x \in X$ there holds

$$\frac{\phi(x, \alpha)}{\alpha} \geq \beta.$$

Bibliographical remarks

Theorem 3.1.6 was published in Kozlowski [5]. The rest of the material has not been published before. Examples 3.2.5 and

3.2.6, however, are based on an example taken from Kozlowski [1]. Some heuristic considerations about the Δ_2 and Δ_2'-conditions were given by Kozlowski and Lewicki [2]. A version of Theorem 3.4.1 for Orlicz spaces may be found in the book of Rolewicz [1, Th.4.2.2] and for Musielak-Orlicz spaces in the book of Musielak [1, Th.13.22]. Similarly, the existence of nontrivial functionals in Orlicz spaces was considered in the book of Rolewicz [1, Th.4.2.1].

4. Special Cases

In this chapter we will present some special cases and examples of function modulars and function semimodulars. In Chapter 1 we already mentioned some of them, presenting examples of modulars. Now we want to give more detailed description of these functionals as well as to present some new examples; we are also able to have a look at some function spaces from the point of view of the theory developed throughout Chapters 2 and 3. Some important special cases are not included in the following chapter because we devote separate chapters to them. Thus, the construction of the function semimodulars for given nonlinear operators, which was mentioned in the Introduction, will be presented in Chapter 6, while the countably modulared spaces and their applications to the theory of summation will be described in Chapter 5. Special cases of function modulars defined on subsets of the n-th Cartesian product of the complex plane C^n as well as some applications to approximation theory and problems of analytic extension will be disscussed in Chapter 7.

4.1 MUSIELAK - ORLICZ SPACES

Musielak-Orlicz spaces may be regarded as modular function spaces. We will recall some basic concepts of the theory of Musielak-Orlicz spaces turning, the reader's attention to its function modular aspects.

Let (X,Σ,μ) be a measure space. Assume μ to be a nonnegative σ-finite measure and denote by $\mathcal{P}$ the δ-ring of all sets of finite measure. A function $\phi : X \times \mathbb{R}^+ \to \mathbb{R}^+$ will be called a ϕ_1-function if and only if ϕ satisfies the following conditions:

(4.1.1) $\phi(x,\cdot)$ is continuous for every $x \in X$;

(4.1.2) $\phi(x,0) = 0$ for every $x \in X$;

(4.1.3) $\phi(x,u) \to \infty$ as $u \to \infty$;

(4.1.4) $\phi(\cdot,u)$ is a measurable, locally integrable function for every $u \geq 0$.

(4.1.5)   There exists a number $\alpha_0 \geq 0$ such that $\phi(\cdot,u) > 0$ ρ-a.e for every $u > \alpha_0$;

If $\alpha_0 = 0$ then ϕ is called a ϕ-function. Let us consider the case of real functions, i.e. $S = \mathbb{R}$. It is easily seen that the functional given by

(4.1.6) $$\rho(f,E) = \int_E \phi\Big(x,|f(x)|\Big)\, d\mu$$

is a function semimodular whenever ϕ is a ϕ_1-function and ρ is a function modular if ϕ is a ϕ-function. The function modular space induced by ρ is called Musielak-Orlicz space (Orlicz space if ϕ does not depend on the first variable) and will be denoted by L^ϕ. It is not hard to see that in this situation we have the following equalities:

(4.1.7) $L_\rho^0 = L_\rho^f = L_0^\phi = \{f \in M(X,\mathbb{R}) \ ; \ \rho(\lambda f) < \infty$ for some $\lambda > 0 \}$;

(4.1.8) $E_\rho = E^\phi = \{f \in M(X,\mathbb{R}) \ ; \ \rho(\lambda f) < \infty$ for every $\lambda > 0 \}$.

The set E^ϕ is usually called a subspace of finite elements.

4.1.9 PROPOSITION. If ρ is a Musielak-Orlicz semimodular then ρ-null sets coincide with the sets of measure zero in the sense of μ.

PROOF. It is evident that every μ-null set is ρ-null as well. To prove the inverse, let us assume that $\rho(\alpha, E) = 0$ for every $\alpha > 0$, which means that

$$\int_E \phi(x,\alpha) \ d\mu = 0 \quad \text{for every } \alpha > \alpha_0.$$

Since ϕ is a positive function this is only possible when

$\mu(E) = 0$ or $\phi(\cdot,\alpha) = 0$ ρ-a.e. for every $\alpha > \alpha_0$ but in view of
(4.1.5) we get $\phi(\cdot,\alpha) > 0$ ρ-a.e. and, consequently, E is a set of
measure μ zero.

The meaning of Proposition 4.1.9 is that we can treat
Musielak-Orlicz spaces as modular function spaces. Following
Musielak [1], we can state another result in which we compare
different ways of defining "Δ_2-type " conditions.

4.1.10 PROPOSITION. Let ϕ be a ϕ-function. For the function
modular ρ induced by ϕ the following two conditions are
equivalent:

(i) ρ satisfies the Δ_2-condition (see Def. 3.1.4);

(ii)   There exists a constant $K > 0$ and a nonnegative,
           integrable function $h : X \to \mathbb{R}^+$ such that
           for every $u \geq 0$ there holds:

$$\phi(x,2u) \leq K \, \phi(x,u) + h(x) \quad \rho\text{-a.e.}$$

Let us recall that a measure μ is called separable if and
only if there exists a countable family of sets $\mathcal{A} \subset \mathcal{P}$ such that to

every A $\in$ $\mathcal{P}$ there corresponds a sequence of sets A_k $\in$ $\mathcal{A}$ with $\mu(A \triangle A_k) \to 0$. By the σ-finiteness of μ and the local integrability of every function of the form $\phi(\cdot,\alpha)$, it is not difficult to prove the next result.

4.1.11 PROPOSITION. For a Musielak-Orlicz modular the following equivalence holds:

(i) μ is separable;

(ii) ρ is separable on $\mathcal{E}$.

Since we have now $S = \mathbb{R}$ and $\mathbb{R}$ is separable, we may reformulate Theorem 3.3.13.

4.1.12 PROPOSITION. Let μ be separable. Then, the following conditions are mutually equivalent:

(a) L^ϕ is separable;

(b) $L^\phi = E^\phi = L_0^\phi$;

(c) L_0^ϕ is a linear subspace of L^ϕ;

(d) ϕ satisfies (4.1.10) (ii);

(e) ρ satisfies Δ_2;

(f)   The modular and the F-norm convergences are equivalent in L^ϕ.

As we can observe, the main facts of the theory of Musielak-Orlicz spaces may be described in terms of the theory of modular function spaces. If one restricted the function ϕ more (e.g. it were assumed to be convex), however, one would obtain results which do not hold in general in modular function spaces (e.g. some theorems on representation of linear functionals). On the other hand, the theory of Musielak-Orlicz spaces does not have good structural properties. For instance, the countably (even finite) modulared space constructed from a sequence of Musielak-Orlicz modulars does not produce in general a Musielak-Orlicz space. In Chapter 5 we will prove that this procedure applied to function semimodulars will produce again a modular function space. To give another example let us consider an increasing, continuous function $\Lambda : [0,\infty] \rightarrow [0,\infty]$ such that $\Lambda(0) = 0$. Let ρ_1 be a function modular with domain M(X,S). Defining then

$$\rho_2(f,E) = \Lambda \left(\rho_1(f,E) \right) \text{ for } E \in \Sigma \text{ and } f \in \mathcal{E},$$

we obtain again a function modular. These function modulars are evidently equivalent which means that $\rho_2(f_n) \rightarrow 0$ if and only if $\rho_1(f_n) \rightarrow 0$. If ρ_1 is a Musielak-Orlicz modular then we do not know whether L_{ρ_2} is an Musielak-Orlicz space. Nevertheless, the modular function space L_{ρ_2} is obviously isomorphic to L_{ρ_1}. The following function Λ is useful:

$$\Lambda(t) = \frac{t}{1+t} \text{ , } t \in [0,\infty].$$

Using as usual the limit convention $\frac{\infty}{1+\infty} = 1$, we get the finite modular

$$\rho_2(f,E) = \Lambda\Big(\rho_1(f,E)\Big).$$

If ρ_1 is a Musielak-Orlicz modular then the notions of the space of finite elements E^ϕ and of Musielak-Orlicz class L_0^ϕ do not have any meaning for the space L_{ρ_2}. It is clear, however, that we may apply our theory and immediately observe that

$$L_{\rho_1} = L_{\rho_2}, \; E_{\rho_1} = E_{\rho_2}, \; L_{\rho_1}^0 = L_{\rho_2}^0.$$

When dealing with applications, the optimal method seems to consist in using the general theory of modular function spaces in order to state corectly the problem and obtain general results and then, if the function semimodulars under consideration are of specific form, apply any particular theory we need. In this way we are able to refine our results and then, using again the theory of modular function spaces, we can give precise answers to the questions stated before. We will use this method throughout last three chapters.

4.2. SOME GENERALIZATIONS

Let us consider now some of the frequently used generalizations of Musielak-Orlicz spaces. Suppose we have a family $\mathcal{M}$ of σ-additive measures on (X,Σ). Let ϕ denote a ϕ_1-function. We may define a function semimodular ρ by the formula:

$$(4.2.1) \qquad \rho(f,E) = \sup_{\mu \in \mathcal{M}} \int_E \phi\Big(x,|f(x)|\Big) \, d\mu.$$

Generally speaking, the functionals of that form are not disjointly additive and, moreover, the classes E_ρ and L_ρ^0 must be defined rather in the form presented in the theory of modular function spaces than in the theory of Musielak-Orlicz spaces, because conditions described by means of finiteness of (4.2.1) usually do not work. Similarly, the condition Δ_2 from the theory of Musielak-Orlicz spaces (condition (ii) from Proposition 4.1.10) must be strengthened by adding a function modular assumption that the set function

$$\sup_{\mu \in \mathcal{M}} \int_{(\cdot)} h \, d\mu$$

is order continuous.

4.2.2 Lorentz type L^P-spaces

As an example of the function semimodular of that type we can mention the following generalization of L^P-space:

$$\rho(f,E) = \sup_{z \in \mathcal{Z}} \int_E |f(x)|^P z(x) \, d\mu,$$

where μ is a fixed measure on X and $\mathcal{Z}$ is a family of nonnegative μ-measurable functions. In particular, we may define

$$\rho(f,E) = \sup_{\tau \in \mathcal{T}} \int_E |f(x)|^P \, d\mu_\tau,$$

where μ is a fixed σ-finite measure on X, $\mathcal{T}$ is a group of all measure preserving transformations $\tau : X \to X$ and μ_τ are given by

$$\mu_\tau(E) = \mu\left(\tau^{-1}(E)\right).$$

4.2.3 Countably modulared spaces

Among other interesting cases let us mention the functionals of the forms

$$\rho_0(f,E) = \sup_{n \in \mathbb{N}} \rho_n(f,E) \quad \text{or} \quad \rho(f,E) = \sum_{n=1}^{\infty} \frac{\rho_n(f,E)}{1+\rho_n(f,E)},$$

where ρ_n are either Musielak-Orlicz semimodulars or function semimodulars given by (4.2.1). These semimodulars will be considered in Chapter 5.

4.2.4 Fenchel-Orlicz spaces

There exist many attempts to create a suitable theory of Orlicz or Musielak-Orlicz spaces for the case of functions taking values in a Banach space S. Let us consider the case of Fenchel-Orlicz spaces introduced by B. Turett in [1]. Fenchel-Orlicz spaces are the modular spaces given by the modular

$$\rho(f) = \int_X \phi\big(f(x)\big) \, d\mu,$$

where the function f takes its values in a Banach space $(S,|\cdot|)$, μ is a σ-finite measure and ϕ is a Young function; this means $\phi : S \to [0,\infty]$ is convex, even, $\phi(0) = 0$ and

$$\lim_{t \to \infty} \phi(tr) = \infty \quad \text{whenever } r \in S \text{ and } r \neq 0.$$

Clearly, this example lacks the monotonicity of modular which is a basic property of function modulars. Moreover, one can imagine that in some interesting vector valued generalizations of Orlicz spaces, the monotonicity of modulars would be a very artificial assumption. This leads us to the question of whether the theory developed throughout the preceding chapters can be applied to the case of nonmonotone functionals which have some similar properties to function modulars. In the next section of this chapter we shall demonstrate that for the convex case the answer can be affirmative.

4.3 NONMONOTONE CONVEX FUNCTION MODULARS

To make the problem precise, let us ask whether it is possible (under some reasonable assumptions) to equip the Banach space S with a norm $\|\cdot\|_\phi$ equivalent to the previous norm $|\cdot|$ such that ρ will be nondecreasing with respect to the norm $\|\cdot\|_\phi$.

Let us make the following definition.

4.3.1 DEFINITION. A functional $\rho : M(X,S) \to [0,\infty]$ is called a nonmonotone convex function modular if and only if

(a1) $\rho(f,\cdot) : \Sigma \to [0,\infty]$ is a σ-subadditive measure for every
        $f \in M(X,S)$;

(a2) $\rho(\cdot,X) : M(X,S) \to [0,\infty]$ is a convex modular;

(a3)   P4, P5 and P6 from the definition of function
        semimodulars are satisfied;

(a4) $cl_{\|\cdot\|_\rho}(\mathcal{S}) = E_\rho \subset L_\rho \subset M(X,S).$

4.3.2 DEFINITION. A nonmonotone convex modular ρ is said to be accurate if there exists a partition (X_n) of X ($X_n \in \mathcal{P}$ are mutually disjoint) such that a modular $\phi : S \to [0,\infty]$ defined by

$$\phi(r) = \sup_{n \in \mathbb{N}} \rho(r1_X, X_n)$$

satisfies:

   (b1) ϕ is continuous at zero in S;

   (b2) $\phi(r) < \infty$ for every $r \in S$;

   (b3) $\phi(\alpha r_1) \le \phi(\alpha r_2)$ for all $\alpha > 0$
            whenever $\phi(r_1) \le \phi(r_2)$, $r_i \in S$.

   (b4) $\rho(f,E) = \sup\{\rho(g,E); g \in \mathcal{S}, \phi(g(x)) \le \phi(f(x))$ in $E\}$,
            where $f \in M(X,S)$ and $E \in \Sigma$.

Observe that, on account of convexity of ϕ, it follows from (b2) that for every $r \in S$ the real function $\lambda \mapsto \phi(\lambda r)$ is continuous.

   Let ρ be an accurate convex function modular; then we can equip the Banach space S with a new norm $\|\cdot\|_\phi$ defined by the formula

$$\|r\|_\phi = \inf\left\{\alpha > 0 \; ; \; \phi\left(\tfrac{r}{\alpha}\right) \le 1\right\}, \, r \in S.$$

Since ϕ is continuous at zero, it follows that the modular space

S_ρ induced by ϕ coincides with S. Note that by (b3), $\|r_1\|_\phi \leq \|r_2\|_\phi$ holds if $\phi(r_1) \leq \phi(r_2)$. For a set $A \in \mathcal{P}$, put

$$\mathcal{C}(A) = \{\ r1_A\ ;\ r \in S\ \}.$$

4.3.3 LEMMA. For every $A \in \mathcal{P}$, the set $\mathcal{C}(A)$ is closed in L_ρ.

PROOF. Let us note that it suffices to prove that $\mathcal{E} \setminus \mathcal{C}(A)$ is open since $\mathcal{E}$ is dense in the closed set E_ρ. Let us fix a function $f \in \mathcal{E} \setminus \mathcal{C}(A)$. Since f is simple, it follows that its range is a finite set $\{r_1, \dots, r_k\} \subset S$. Let $\delta > 0$ be such that $\|r_1 - r_i\|_\phi > 2\delta$ for $i = 2, \dots, k$. Thus, for given $r \in S$ there holds either $\|r - r_1\|_\phi > \delta$ or $\|r - r_i\|_\phi > \delta$ for $i = 2, \dots, k$. Hence,

$$\|r - f(x)\|_\phi > \delta \text{ for all } x \in E$$

or

$$\|r - f(x)\|_\phi > \delta \text{ for all } x \in F,$$

where

$$E = f^{-1}\big(\{r_1\}\big) \text{ and } F = f^{-1}\big(\{\ r_2, \dots, r_k\}\big).$$

Let us suppose, for instance, that the first possibility takes place. Hence,

$$\inf\Big\{\alpha > 0\ ;\ \phi\Big(\frac{r - f(x)}{\alpha}\Big) \leq 1\Big\} > \delta \text{ for all } x \in E.$$

Therefore,

$$\phi\Big(\frac{r - f(x)}{\delta}\Big) > 1 \text{ for all } x \in E.$$

Let $t \in \phi^{-1}\big(\{1\}\big)$; then

$$\left\{ x \in X; \; \phi\!\left(\frac{r1_A(x) - f(x)}{\delta} \right) > 1 \right\}$$

$$= \left\{ x \in X; \; \phi\!\left(\frac{r1_A(x) - f(x)}{\delta} \right) > \phi(t) \right\} \supset E.$$

By (a1), (a2) and (b4) we obtain

$$\rho\!\left(\frac{r1_A - f}{\delta} \right) \geq \rho\!\left(\frac{r1_A - f}{\delta}, \; E \right) \geq \rho(t1_E, E) > 0.$$

Denoting $k_E = \rho(t1_E, E)$, we observe that two possibilities arise:

(1)   if $k_E \geq 1$ then $|\|r1_A - f\|_\rho > \delta$;

(2)   if $k_E < 1$ then

$$\rho\!\left(\frac{r1_A - f}{\delta k_E} \right) \geq \rho\!\left(\frac{r1_A - f}{\delta} \right) > 1 \text{ for all } x \in E.$$

Since the same can be done for F,

$$\|r1_A - f\|_\rho > \min\{\delta, \; \delta k_E, \; \delta k_F\} > 0,$$

where

$$k_F = \rho(t1_F, F).$$

Consequently, $\mathcal{E} \setminus \mathcal{C}(A)$ is an open subset of L_ρ.

4.3.4 LEMMA. $(S,\|\cdot\|_\phi)$ is complete.

PROOF. Let (r_n) be a Cauchy sequence in $(S,\|\cdot\|_\phi)$. Let $A = X_k$ for a certain $k \in \mathbb{N}$. Denoting $f_n = r_n 1_A \in \mathcal{C}(A)$, we observe that

$$\rho\Big(\alpha(f_n - f_m)\Big) = \phi\Big(\alpha(r_n - r_m)\Big) \to 0 \quad \text{as } n,\, m \to \infty.$$

L_ρ is complete; therefore, there exists a function $f \in L_\rho$ such that $\|f_n - f\|_\rho \to 0$. By Lemma 4.3.3, $\mathcal{C}(A)$ is closed in L_ρ, and therefore, $f = r1_A$ for a certain $r \in S$. Finally,

$$\phi\Big(\alpha(r_n - r)\Big) = \rho\Big(\alpha(f_n - f)\Big) \to 0,$$

which completes the proof.

4.3.5 THEOREM. The norm $\|\cdot\|_\phi$ is equivalent to $|\cdot|$.

PROOF. Let us consider the identity map $I : (S,|\cdot|) \to (S,\|\cdot\|_\phi)$. We will prove that I is continuous. Indeed, let $|r_n| \to 0$; then $|\alpha r_n| \to 0$ for every $\alpha > 0$, which implies $\|r_n\|_\phi \to 0$. Since I is a linear isomorphism and both spaces $(S,|\cdot|)$ and $(S,\|\cdot\|_\phi)$ are complete, then by the Open Mapping Theorem, they are isomorphic as Banach spaces. This completes the proof.

As a consequences of properties (b2) and (b3) we obtain the next result.

4.3.6 PROPOSITION. $\|r_1\|_\phi \leq \|r_2\|_\phi$ iff $\phi(r_1) \leq \phi(r_2)$.

PROOF. Since $\phi(r_1) \leq \phi(r_2)$ implies $\|r_1\|_\phi \leq \|r_2\|_\phi$, it suffices to prove that $\phi(r_1) = \phi(r_2)$ whenever $\|r_1\|_\phi = \|r_2\|_\phi$. Let

$$\alpha = \|r_1\|_\phi = \|r_2\|_\phi.$$

By the continuity of the function $\lambda \mapsto \phi(\lambda r)$ we conclude that

$$\phi\left(\frac{r_1}{\alpha}\right) = \phi\left(\frac{r_2}{\alpha}\right) = 1$$

and by (b3)

$$\phi(r_1) = \phi(r_2).$$

The following theorem is an immediate consequence of Theorem 4.3.6, Proposition 4.3.5 and the property (b4).

4.3.7 THEOREM. Let ρ be an accurate nonmonotone convex function modular, then ρ is a function modular which is monotone with respect to the equivalent norm $\|\cdot\|_\phi$, i.e.

(i) $\rho(f,E) = \sup\{\rho(g,E); g \in \mathcal{E}, \|g(x)\|_\phi \leq \|f(x)\|_\phi$ in $E\}$, which also implies that

(ii) $\rho(g,E) \leq \rho(f,E)$ whenever f, g $\in$ M(X,S) and

$\|g(x)\|_\phi \leq \|f(x)\|_\phi$ for x $\in$ E.

4.4 NONLINEAR OPERATOR VALUED MEASURES

Let F be a Banach space. By N(S,F) we denote the space of all mappings U : S $\rightarrow$ F such that U(0) = 0 and U are uniformly continuous on bounded subsets of S. A set function μ : $\mathcal{P}$ $\rightarrow$ N(S,F) is said to be an operator valued measure if μ has the following properties:

(4.4.1) $\mu(0) = 0$;

(4.4.2) μ is countably additive in the pointwise sense;

(4.4.3) $\lim\limits_{\delta \to 0} sv_\delta (\mu,\alpha,E) = 0$ for every E $\in$ $\mathcal{P}$ and $\alpha > 0$;

(4.4.4) $\widetilde{\mu}(\cdot)r$ is order continuous for every r $\in$ S.

Let us recall two quantities that have been used above:

$$sv_\delta(\mu,\alpha,E) = \sup\left\{\left\|\sum_{i=1}^{n}[\mu(E_i)r_i - \mu(E_i)r_i']\right\|; \bigcup_{i=1}^{n} E_i \subset E, E_i \in \mathcal{P},\right.$$

$$\left. |r_i|, |r_i'| \leq \alpha, |r_i - r_i'| \leq \delta, 1 \leq i \leq n, n \in \mathbb{N}\right\}.$$

The other one is defined as follows:

$$\widetilde{\mu}(E)r = \sup\{\|\mu(A)r\| \; ; \; A \subset E, A \in \mathcal{P}\}.$$

For a $\mathcal{P}$-simple function

$$f = \sum_{i=1}^{n} r_i 1_{E_i} \text{ where } r_i \in S, E_i \in \mathcal{P},$$

the integral was defined as

$$\int_E f \, d\mu = \sum_{i=1}^{n} \mu(E \cap E_i) r_i, \; E \in \Sigma.$$

Then, the domain of integration can be extended to the space $\mathcal{M}(\mu)$ of all measurable f such that there exists a sequence of simple functions (f_n) converging everywhere to f for which the integrals

$$\left(\int_{(\cdot)} f_n \, d\mu \right)$$

are uniformly countably additive on Σ. Let us assume additionally that there exists $\alpha > 0$ such that

(4.4.5)   if $\displaystyle\int_E f \, d\mu = 0$ for all $f \in \mathcal{E}$ such that $|f(x)| \leq \alpha$ in E,

then $\displaystyle\int_E g \, d\mu = 0$ for all $g \in \mathcal{E}.$

Define then

$$\rho(f,E) = \sup\Big\{\|\int_E g\,d\mu\,\|\ ;\ g\in \mathcal{S},\ |g(x)|\le |f(x)|\ \text{for every }x\in E\Big\}.$$

The functional ρ thus obtained is a function modular. Properties P1, P2 and P3 follow immediately from the definition of ρ while (4.4.3) gives P4, (4.4.5) implies P5 and the property P6 is a consequence of (4.4.4).

The integral introduced above may be regarded as a nonlinear operator defined in $\mathcal{M}(\mu)$ with values in the Banach space F. It may be proved that $L_\rho^0 \subset \mathcal{M}(\mu)$ and that the integral is continuous in E_ρ. These facts correspond to the results from Chapter 6 where we will prove them in the more general setting of nonlinear disjointly additive operators. We will state there that to every operator T from a large class of nonlinear operators we may define, in a similar way, a function semimodular ρ in such a way that T is continuous in E_ρ.

Bibliographical remarks

The best review of the theory of Musielak-Orlicz spaces may be found in the book of J. Musielak [1], where they are called generalized Orlicz spaces. First results concerning Orlicz spaces *induced by a convex function ϕ independent of the parameter t were obtained by W. Orlicz [1], [2] and by Z. Birnbaum and W. Orlicz [1] in early 1930's. The book of Krasnosel'skii and Rutickii [1] gives the exposition of that classical theory. The idea of Musielak-Orlicz modulars mentioned earlier by Nakano in his book of 1950 [1], was further developed in 1959 by Musielak and Orlicz in their papers [1] and [2]. This theory has been extensively elaborated for almost three decades. For instance, many mathematicians obtained characterizations of geometrical properties of Musielak-Orlicz spaces (see e.g. papers of Hudzik [1], Kamińska [3] and Turett [1]). Interesting results on interpolation in such function spaces were considered by Peetre [1], Hudzik, Musielak, Urbański [1], [2], [3] and by Krbec [1]. The generalized Orlicz spaces induced by families of measures were considered in general by Kamińska in her thesis [1] and then in a revised form in a joint paper by Drewnowski and Kamińska [1]. In a less general setting the spaces of that type were considered by Rosenberg [1], Musielak and Waszak [2]. Let us also mention that a space of similar type was used by Szankowski [1] to construct a Banach lattice without the approximation property. The theory of Fenchel-Orlicz spaces was developed by Turett in [2]. This theory inspired Kozlowski [5] to consider the nonmonotone convex function modulars. The linear operator valued measures were investigated by Dobrakov in a series of papers starting from [1] and [2]. Integration of bounded functions with respect to nonlinear operator valued measures* *was considered by many authors in connection with representation of orthogonally additive operators (see Batt [1], Friedman and Tong [1]). Then, this integration was extended to the larger class $\mathcal{M}_b(\mu)$ in papers [2], [3] of Kozlowski and Szczypiński. For the basic exposition of this theory we refer the reader to the paper of Szczypiński [1]; the continuity of the integral was proved in Kozlowski [5].*

5. Countably Modulared Function Spaces

Let (X, Σ, μ) be a finite measure space and let (p_k) be a nondecreasing sequence of numbers with $p_k \geq 1$ for all $k \in \mathbb{N}$. We can introduce a function norm

$$\|f\|_\Sigma = \sup_{k \in \mathbb{N}} \|f\|_{p_k},$$

where $\|\cdot\|_{p_k}$ stands for the usual norm in the space L^{p_k}. Denoting by L_Σ the Banach function space induced by the norm $\|\cdot\|_\Sigma$, we can observe that L_Σ is equal to the intersection of all spaces L^{p_k} if and only if the sequence (p_k) is constant beginning from some index $i \in \mathbb{N}$, i.e. there exists $i \in \mathbb{N}$ such that $L^{p_i} = L_\Sigma = L^{p_k}$ for $k \geq i$. We may ask whether the same result holds for functions summable in the sense of Césaro, that is, when the Lebesgue norms are replaced by

$$\|f\|_{p_k} = \limsup_{\tau \to \infty} \frac{1}{\tau} \int_0^\tau |f(t)|^{p_k} dt.$$

It will turn out that the answer is affirmative. Using the theory of modular function spaces, we are going to prove results of that type for a large class of function spaces. Section 5.1 is preliminary. The main result, Theorem 5.2.19, will be proved in

Section 5.2 while in Section 5.3 we will present a list of function spaces for which Theorem 5.2.19 holds.

5.1 GENERAL RESULTS

Let (ρ_n) be a sequence of function pseudomodulars. Let us assume additionally that f(x)=0 for all $x \in A$, if $A \in \Sigma$ and $\rho_n(f,A)=0$ for all $n \in \mathbb{N}$, $A \in \Sigma$. Let us define the following functionals:

$$\rho(f,A) = \sum_{n=1}^{\infty} 2^{-n} \frac{\rho_n(f,A)}{1+\rho_n(f,A)} \ , \ \ \rho_0(f,A) = \sup_{n \in \mathbb{N}} \rho_n(f,A), \ \ A \in \Sigma.$$

We used the convention $\frac{\infty}{1+\infty} = 1$ here.

It is evident that both ρ and ρ_0 are modulars in the space of all measurable functions. The corresponding spaces L_ρ and L_{ρ_0} are called the countably modulared space and the uniformly countably modulared space, respectively. Observe that

$$L_\rho = \bigcap_{i=1}^{\infty} L_{\rho_i} \ \text{and} \ L_{\rho_0} \subset L_\rho,$$

this embedding being continuous both with respect to modular and F-norm convergence. We are, however, interested in the question of whether ρ and ρ_0 are the function modulars.

5.1.1 THEOREM. The modular ρ introduced above has properties P1 through P6.

PROOF. It follows immediately from properties of ρ_n and of the real function $t \mapsto \frac{t}{1+t}$ ($t \in [0,\infty]$) that ρ satisfies the conditions P1, P2, P3 and P5. In order to prove P4 and P6 we have to check first that

$$(5.1.2) \qquad \rho(\alpha,A) \leq \sum_{n=1}^{\infty} 2^{-n} \frac{\rho_n(\alpha,A)}{1 + \rho_n(\alpha,A)}$$

Indeed,

$$\rho(\alpha,E) = \sup\{\rho(g,A); \ g \in \mathcal{E}, \ |g(x)| \leq \alpha \text{ for all } x \in A\}$$

$$= \sup\left\{\sum_{n=1}^{\infty} 2^{-n} \frac{\rho_n(g,A)}{1+\rho_n(g,A)} \ ; \ g \in \mathcal{E}, \ |g(x)| \leq \alpha \text{ for all } x \in A\right\}$$

$$\leq \sum_{n=1}^{\infty} 2^{-n} \sup\left\{ \frac{\rho_n(g,A)}{1+\rho_n(g,A)} \ ; \ g \in \mathcal{E}, \ |g(x)| \leq \alpha \text{ for all } x \in A\right\}$$

$$\leq \sum_{n=1}^{\infty} 2^{-n} \frac{\rho_n(\alpha,A)}{1 + \rho_n(\alpha,A)} \ .$$

The last inequality is due to the fact that the function $t \mapsto \frac{t}{1+t}$ is increasing. Since ρ_n is a function pseudomodular then for each n separately $\rho_n(\alpha,A) \to 0$ as $\alpha \to 0$. Clearly,

$$\sum_{n=1}^{\infty} 2^{-n} \frac{\rho_n(\alpha,A)}{1 + \rho_n(\alpha,A)} \to 0 \ \text{ as } \alpha \to 0.$$

Hence, by (5.1.2) we obtain that $\rho(\alpha,E) \to 0$ as $\alpha \to 0$. This completes the proof of P4; the property P6 may be proved similarly.

The next theorem deals with uniformly countably modulared spaces.

5.1.3 THEOREM. ρ_0 is a function modular if and only if the following conditions are simultanously satisfied:

(a) $\displaystyle\sup_{k\in\mathbb{N}} \rho_k(\alpha,A_n) \to 0$ for all $\alpha > 0$, $A_n \in \mathcal{P}$, $A_n \downarrow \emptyset$,

(b) $\displaystyle\sup_{k\in\mathbb{N}} \rho_k(\alpha,A) \to 0$ as $\alpha \to 0$, $A \in \mathcal{P}$.

PROOF. It is clear that ρ_0 satisfies conditions P1, P2, P3 and P5, while P4 and P6 are implied by (a) and (b).

An answer to the question, under which conditions are both spaces L_ρ and L_{ρ_0} equal, may be formulated as follows:

5.1.4 PROPOSITION. $L_\rho = L_{\rho_0}$ if and only if condition (a) from Theorem 5.1.3 holds and also

(b') $\displaystyle\sup_{k\in\mathbb{N}} \rho_k(\lambda_n f,A) \to 0$ as $\lambda_n \to 0^+$, $f \in L_\rho$, $A \in \mathcal{P}$.

Let us observe that (b') implies (b). Therefore, the following theorem holds:

5.1.5 THEOREM. If $L_\rho = L_{\rho_0}$ then ρ_0 is a function modular.

It is obvious that ρ_0 need not be a function modulars. Consider for instance the sequence

$$\rho_n(f,A) = \int_A |f|^n dm;$$

ρ_0 induced by this sequence does not satisfy the condition P6. On the other hand there are examples of sequences (ρ_k) such that ρ_0 is a function modular while $L_\rho \neq L_{\rho_0}$.

5.1.6 EXAMPLE. Let m denote the Lebesgue measure in $X = [0,1]$ and let $\phi_1(u) = |u|$. For $k \geq 2$ put

$$\phi_k(u) = \begin{cases} 0, & x \in [0,1], \\ 2^{(n-1)^2}, & x \in [n-1,n), \ 2 \leq n \leq k, \\ 2^{k^2}, & x \geq k. \end{cases}$$

Let

$$\rho_k(f,A) = \int_A \phi_k(f(x)) \, dm(x) \quad \text{and} \quad \rho_0(f,A) = \sup_{k \in \mathbb{N}} \rho_k(f,A).$$

Observe that

$$\rho_0(\alpha,A) = m(A) \sup_{k \in \mathbb{N}} \phi_k(\alpha) \leq m(A) \, 2^{(n-1)^2} \quad \text{if } \alpha \in [n-1,n) \ .$$

Thus (a) and (b) from (5.1.3) hold and consequently ρ_0 is a function modular. Put $f = \sum n 1_{I_n}$, where $I_n \subset X$ are mutually disjoint and $m(I_n) = 2^{-n}$. We claim that $f \in L_\rho = \bigcap_{k=1}^{\infty} L_{\rho_k}$. Given a sequence $0 \leq \gamma_m \to 0$, let us choose a subsequence (λ_j) of (γ_n) such that $\lambda_j \leq j^{-1}$. We have then

$$\rho_1(\lambda_j f) = \lambda_j \sum_{n=1}^{\infty} n \, m(I_n) \leq c \, \lambda_j \to 0, \text{ where } c = \sum_{n=1}^{\infty} n \, 2^{-n} < \infty.$$

For $k \geq 2$ there holds:

$$\rho_k(\lambda_j f) \leq \rho_k(f j^{-1}) \leq 2^{k^2} \sum_{n=j}^{\infty} m(I_n) = 2^{k^2} \sum_{n=j}^{\infty} 2^{-n} \to 0 \text{ as } j \to \infty.$$

We used here the fact that for $n < j$ and $x \in I_n$ there holds

$$\frac{f(x)}{j} = \frac{n}{j} < 1$$

and consequently

$$\phi_k\left(\frac{f(x)}{j}\right) = 0.$$

On the other hand, for an arbitrary $j \in \mathbb{N}$, and for $n \geq kj$, $k \geq 2$, $x \in I_n$, we have

$$\frac{f(x)}{j} = \frac{n}{j} \geq k.$$

Thus, $\phi_k\left(\dfrac{f(x)}{j}\right) = 2^{k^2}$ and

$$\rho_0\left(\tfrac{1}{j}\cdot f\right) = \sup_{k \in \mathbb{N}} \rho_k\left(\tfrac{f}{j}\right) \geq \sup_{k \in \mathbb{N}} 2^{k^2} \sum_{n=kj}^{\infty} m(I_n)$$

$$= \sup_{k \in \mathbb{N}} 2^{k^2} \sum_{n=kj}^{\infty} 2^{-n} = \sup_{k \in \mathbb{N}} 2^{k^2-kj+1} = \infty.$$

Finally, $\rho_0\left(\dfrac{1}{j}\cdot f\right)$ does not converge to zero as $j \to \infty$, i.e. f does not belong to L_{ρ_0}.

5.2 (ϕ,a) -SUMMABLE FUNCTIONS

To the end of this section we will restrict our consideration to some more special modulars of the form:

$$(5.2.1) \qquad \rho(f) = \sup_{\tau \in \mathcal{T}} \int_{t_0}^{\infty} \phi\Big(|f(t)|\Big)\, d\mu_\tau$$

or

$$(5.2.2) \qquad \rho(f) = \lim_{\tau \to \infty} \sup \int_{t_0}^{\infty} \phi\Big(|f(t)|\Big)\, d\mu_\tau,$$

where $\phi : \mathbb{R}^+ \to \mathbb{R}^+$ is a ϕ-function (cf. Section 4.1) which does not depend on the first variable and $\{\mu_\tau\}_{\tau \in \mathcal{T}}$ is a family of uniformly bounded measures on $\mathbb{R}^+$, i.e. there exists $K < \infty$ such that

$$\sup_{\tau \in \mathcal{T}} \mu_\tau(X) = K.$$

Though it is possible to define the modular of type (5.2.2) for an abstract set of of indices $\mathcal{T}$, we restrict our consideration to the set $\mathcal{T}$ of the form $\mathcal{T} = [\tau_0, \infty)$, where $\tau_0 \geq 0$. Observe that for the modulars ρ of the form given by (5.2.1), the topological nature of the set $\mathcal{T}$ does not matter at all.

The theory of countably modulared spaces for the case of such function modulars was developed by J. Musielak and A. Waszak in the late 1960's (cf. the bibliographical remark at the end of the chapter). The reader interested in this subject may find more references in the book of Musielak [1].

In particular, we will be interested in the cases:

(a) $X = [t_0, \infty)$, $\mathcal{T} = [\tau_0, \infty)$, $t_0, \tau_0 > 0$, and

$$\mu_\tau(A) = \int_A a(t, \tau) dt$$

for $\tau \in \mathcal{T}$, where $a : X \times \mathcal{T} \to \mathbb{R}$ is a Lebesgue measurable function in its first variable. In this situation the modulars (5.2.1) and (5.2.2) take the forms:

(5.2.3) $$\rho(f) = \sup_{\tau \in \mathcal{T}} \int_{t_0}^{\infty} a(t, \tau) \, \phi\Big(|f(t)|\Big) \, dt,$$

(5.2.4) $$\rho(f) = \lim_{\tau \to \infty} \sup \int_{t_0}^{\infty} a(t,\tau)\, \phi\Big(|f(t)|\Big)\, dt.$$

(b) $$X = \mathbb{N},\ \mathcal{T} = \mathbb{N},\ t_0 = \tau_0 = 1 \text{ and}$$

$$\mu_n(A) = \sum_{i \in A} a_{n,i}$$

for every $A \subset \mathbb{N}$ ($\mu_n(\emptyset) = 0$ by definition), where $a=(a_{n,i})$ is an infinite matrix of nonnegative numbers with no column consisting of zeros only. In this case we define

(5.2.5) $$\rho\Big((t_i)\Big) = \sup_{n \in \mathbb{N}} \sum_{i=1}^{\infty} a_{n,i}\, \phi(t_i),$$

(5.2.6) $$\rho\Big((t_i)\Big) = \lim_{n \to \infty} \sup \sum_{i=1}^{\infty} a_{n,i}\, \phi(t_i).$$

It is easy to check that both formulas (5.2.1) and (5.2.2) define the function modulars in the space $M(X)$ of all measurable functions $f : X \to \mathbb{R}$.

5.2.7 DEFINITION. A ϕ-function ϕ is said to be regular if and only if to every $\gamma > 0$ there exist positive numbers u' and α'

such that

(5.2.8) $\phi(\alpha' u) \leq \gamma \cdot \phi(u)$ for all $u \geq u'$.

For instance, every s-convex ϕ-function is regular. A function $\phi(u) = \ln(1 + u)$ for $u > 0$ is an example of a regular ϕ-function which is not s-convex.

5.2.9 PROPOSITION. Let a function modular ρ be given by (5.2.1) or (5.2.2) and let ϕ be a regular ϕ-function. If there exists $\lambda > 0$ such that $\rho(\lambda f) < \infty$ then $f \in L_\rho$.

PROOF. We present the proof for the case of the modular ρ given by (5.2.1). The other version can be proved in the same way. Let us fix a number $\epsilon > 0$ and a measurable function f with $\rho(\lambda f) < \infty$ for a given $\lambda > 0$. Let us put $\gamma = \frac{\epsilon}{2\rho(\lambda f)}$ and pick up the numbers u', $\alpha' > 0$ from (5.2.8), and define two sets A and B by $A = \{t \in X; \lambda|f(t)| \geq u'\}$, $B = X \setminus A$. Let $\lambda_n \to 0^+$. We have to prove that $\rho(\lambda_n f) \to 0$ as $n \to \infty$. Take $n_0 \in \mathbb{N}$ so large that $\lambda_n < \alpha'\lambda$ for every $n \geq n_0$. For $n \geq n_0$ we have then:

$$\rho(\lambda_n f) \leq \sup_{\tau \in \mathcal{T}} \int_A \phi\left(\frac{\lambda_n}{\lambda} \lambda|f(t)|\right) d\mu_\tau + \sup_{\tau \in \mathcal{T}} \int_B \phi\left(\frac{\lambda_n}{\lambda} \lambda|(f(t)|\right) d\mu_\tau$$

$$\leq \sup_{\tau \in \mathcal{T}} \int_A \phi\Big(\alpha'\lambda|f(t)|\Big) \, d\mu_\tau + \phi\Big(\frac{\lambda_n}{\lambda} u'\Big) \sup_{\tau \in \mathcal{T}} \mu_\tau(B)$$

$$\leq \frac{\epsilon}{2\rho(\lambda f)} \rho(\lambda f) + K \, \phi\Big(\frac{\lambda_n}{\lambda} \cdot u'\Big) \leq \epsilon$$

for $n \geq n_1 \geq n_0$, where $n_1 \geq n_0$ has been chosen such that

$$\phi\Big(\frac{\lambda_n}{\lambda} \cdot u'\Big) \leq \frac{\epsilon}{2K} \text{ for } n \geq n_1.$$

Recall that ϕ is continuous at zero and $\lambda_n \to 0$.

Either of the modulars (5.2.1) or (5.2.2) induces a particular method of summability (the relation to the classical methods will be disscussed later in this section). Namely, we make the following definition.

5.2.10 DEFINITION. We say that a Σ-measurable function $f : X \to \mathbb{R}$ is (ϕ,a)-summable (resp. topologically (ϕ,a)-summable) if there exists $\lambda > 0$ such that

$$\sup_{\tau \in \mathcal{T}} \int_{t_0}^{\infty} a(\tau,t) \, \phi\Big(\lambda|f(t)|\Big) \, dt < \infty$$

$$\Big(\text{resp. } \limsup_{\tau \to \infty} \int_{t_0}^{\infty} a(\tau,t) \, \phi\Big(\lambda|f(t)|\Big) \, dt < \infty\Big).$$

In the discrete case we have the following formulas:

$$\sup_{n \in \mathbb{N}} \sum_{i=1}^{\infty} a_{n,i} \, \phi(\lambda t_i) < \infty$$

$$\left(\text{resp. } \limsup_{n \to \infty} \sum_{i=1}^{\infty} a_{n,i} \, \phi(\lambda t_i) < \infty \right).$$

If ϕ is regular, then by Proposition 5.2.9, (ϕ,a)-summability (resp. topological (ϕ,a)-summability) is equivalent to being a member of the function modular space L_ρ where ρ is given by (5.2.1) (resp. by (5.2.2)).

Suppose now that we have a sequence (ϕ_k) of regular ϕ-functions. It may happen that a given function $f : X \to \mathbb{R}$ is (ϕ,a)-summable for every ϕ_k separately, which means, in view of Proposition 5.2.9, that for every sequence $\lambda_n \to 0$ there holds $\rho_k(\lambda_n f) \to 0$ as $n \to \infty$. By ρ_k we denote the function modular given by (5.2.1) or (5.2.2) with ϕ replaced by ϕ_k.

The problem mentioned at the beginning of this chapter may be stated in the form: is the function f uniformly (ϕ_k,a)-summable, i.e. is it true that

$$\sup_{k \in \mathbb{N}} \rho_k(\lambda_n f) \to 0 \text{ as } \lambda_n \to 0 \ ?$$

We can reformulate this problem in the language of countably modulared function spaces. Indeed, put

$$\rho(f) = \sum_{k=1}^{\infty} 2^{-k} \frac{\rho_k(f)}{1+\rho_k(f)} \quad \text{and} \quad \rho_0(f) = \sup_{k \in \mathbb{N}} \rho_k(f).$$

We may reduce our problem to the question already disscussed at the beginning of this chapter: when is $L_\rho = L_{\rho_0}$? It will turn out that under some reasonable assumptions the condition $L_\rho = L_{\rho_0}$ is equivalent to the fact that we deal only with a finite number of different modular function spaces. The meaning of the latter is as follows: the equivalence between (ϕ_k,a)-summability for every k and uniform (ϕ_k,a)-summability obviously holds for any finite number of methods but one should not expect any interesting result for an infinite number of methods.

Before we state this result (Theorem 5.2.19) we need some more definitions. Examples of some classical methods of summability and their relationship to Theorem 5.2.19 will be given in the next section.

5.2.11 DEFINITION. A sequence of ϕ-functions $\phi_k : X \to \mathbb{R}^+$ will be called a normal sequence of ϕ-functions if and only if:

(5.2.12)   Every ϕ_k is a regular ϕ-function,

(5.2.13) (ϕ_k) are equicontinuous at zero,

(5.2.14)   The sequence (ϕ_k) is essentially increasing, i.e. to every index $n \in \mathbb{N}$ there correspond three positive numbers λ_n, β_n, v_n such that for every $k \geq n$ and all $u \geq v_n$ there holds $\phi_n(\lambda_n u) \leq \beta_n \phi_k(u)$.

The restrictions imposed on (ϕ_k) in this definition may look very rigid; they are, however, quite natural and describe simply some kind of regular behavior of the functions (ϕ_k) at zero and infinity while (5.2.14) defines a certain monotonicity of this sequence.

The following sequence of ϕ-functions may serve as a good example. Let us put

$$\phi_k(u) = [\phi(\gamma_k u]^{p(k)},$$

where ϕ is an arbitrary regular ϕ-function, (γ_n) and (M_n) are sequences of positive numbers such that

$$\frac{\gamma_n}{\gamma_k} \leq M_n \quad \text{for every} \ k \geq n,$$

while $p(k) \geq 1$ and $p(k + 1) \geq p(k)$ for all $k \in \mathbb{N}$. Certainly, (5.2.12), (5.2.13), (5.2.14) are satisfied for $\phi_k(u) = |u|^{p_k}$, $1 \leq p_1 \leq p_2 \leq \cdots$.

The following property of a sequence of ϕ-functions will be crucial for our further cosiderations.

5.2.15 DEFINITION. We say that the sequence (ϕ_k) of ϕ-functions is essentially constant if there exist three positive numbers K_0, ω, u_0 and an index $i_0 \in \mathbb{N}$ such that for every natural $i \geq i_0$ the following inequality

$$(5.2.16) \qquad \phi_i(\omega u) \leq K_0 \phi_{i_0}(u)$$

is satisfied for every $u \geq u_0$.

5.2.17 REMARK. Observe that if (ϕ_k) is an essentially constant and normal sequence of ϕ-functions then by (5.2.14) and (5.2.16) there exist $i_0 \in \mathbb{N}$ and four positive numbers α, β, δ, u_0 such that for any $i \geq i_0$ we have

$$(5.2.17.a) \quad \delta \phi_{i_0}(\beta u) \leq \phi_i(u) \leq \alpha \phi_{i_0}(u) \quad \text{for all } u \geq u_0.$$

From the left inequality we get by standard arguments of the theory of Orlicz spaces (see e.g. Kufner et al. [1], Th. 3.17.1) that

$$\int_{t_0}^{\infty} \phi_{i_0}\big(\beta|f(x)|\big) \, d\mu_\tau \leq \frac{1}{\beta}\left(\phi_{i_0}(u_0)\,\mu_\tau(X) + \int_{t_0}^{\infty} \phi_i\big(|f(x)|\big)\,d\mu_\tau\right)$$

holds for every measurable function f and every $\tau \in \mathcal{T}$. By the uniform boundedness of measures μ_τ we conclude then that

$$\rho_{i_0}(\beta f) = \sup_{\tau \in \mathcal{T}} \int_{t_0}^{\infty} \phi_{i_0}\big(|\beta f(t)|\big)\,d\mu_\tau \leq \frac{1}{\beta}\big(\phi_{i_0}(u_0)\,K + \rho_i(f)\big).$$

If $f \in L_{\rho_i}$ then $\rho_i(\lambda f) < \infty$ for some $\lambda > 0$ and in view of the above, $\rho_{i_0}(\eta f) < \infty$ for some $\eta > 0$, which via, Proposition 5.2.9, implies that $f \in L_{\rho_{i_0}}$. The other implication, $L_{\rho_{i_0}} \subset L_{\rho_i}$, can be proved similarly using the right-hand inequality in (5.2.17.a). The equality $L_{\rho_i} = L_{\rho_{i_0}}$ means that only the first $i_{i_0} - 1$ spaces may be different.

Let us introduce two properties of a family of measures (μ_τ).

5.2.18   DEFINITION.   The   family (μ_τ) will   be   called equisplittable (topologically equisplittable) if there exists a number $\eta > 0$ such that for every sequence of numbers

$$\epsilon_k \leq \eta, \quad \epsilon_k \downarrow 0, \quad \frac{\epsilon_{k+1}}{\epsilon_k} \leq \frac{1}{2}$$

there exist constants $M \geq m > 0$ and a sequence of pairwise disjoint sets $A_k \in \Sigma$ such that

$$m\epsilon_k \leq \sup_{\tau \in \mathcal{J}} \mu_\tau(A_k) \leq M\epsilon_k, \quad k = 1, 2, \ldots .$$

$$\left(m\epsilon_k \leq \limsup_{\tau \to \infty} \mu_\tau(A_k) \leq M\epsilon_k, \quad k = 1, 2, \ldots \right).$$

For some examples and characterizations of such families of measures the reader is referred to the last section of this chapter. Finally, we are able to state our main result.

5.2.19 THEOREM. Let (μ_τ) be a uniformly bounded, equisplittable family of measures. Suppose (ϕ_k) is a normal family of ϕ-functions. Let

$$\rho(f,A) = \sum_{n=1}^{\infty} 2^{-n} \frac{\rho_n(f,A)}{1+\rho_n(f,A)}, \qquad \rho_0(f,A) = \sup_{n \in \mathbb{N}} \rho_n(f,A),$$

where ρ_n denotes the function modular given by either of the formulas (5.2.1) or (5.2.2). Then the following conditions are equivalent:

(5.2.20) $\quad L_{\rho_0} = L_\rho,$

(5.2.21) $\quad (\phi_k)$ is essentially constant.

PROOF. (5.2.21) $\Rightarrow$ (5.2.20) First, we will show that the fact of being essentialy constant is equivalent to the following condition:

(5.2.22) There exists $i_0 \in \mathbb{N}$ and $\omega' > 0$ such that to every $u' > 0$ there corresponds $K'_0 > 0$ for which

$$\phi_i(u) \le K'_0 \phi_{i_0}(\omega' u)$$

for all $u \ge u'$ and $i \ge i_0$.

Clearly, (5.2.22) implies that (ϕ_k) is essentialy constant. Conversely, suppose (ϕ_k) are essentially constant. We get, therefore, numbers ω, K_0, i_0, u_0 such that

$$\phi_i(\omega u) \le K_0 \phi_{i_0}(u)$$

for every $u \ge u_0$, $i \ge i_0$. Put $\omega' = \max\left(1, \frac{1}{\omega}\right)$ and fix arbitrary $u' > 0$ such that $\phi_{i_0}(u') > 0$. Define

$$K'_0 = \max\left(K_0, \frac{K_0\phi_{i_0}(u_0)}{\phi_{i_0}(u')}\right).$$

Suppose that $u' \ge \omega u_0$ and $u \ge u'$. Hence $\frac{u}{\omega} \ge u_0$ and

$$\phi_i(u) \le \phi_i\left(\omega \cdot \frac{u}{\omega}\right) \le K_0\phi_i\left(\frac{u}{\omega}\right) \le K'_0\phi_i(\omega'u).$$

If $0 < u' < \omega u_0$, $u \ge u'$ and $u \ge \omega u_0$ we obtain

$$\phi_i(u) \le K_0\phi_i\left(\frac{u}{\omega}\right) \le K'_0\phi_i(\omega'u)$$

as previously. Suppose now that $u' \le u \le \omega u_0$. We get

$$\phi_i(u) \le \phi_i(\omega u_0) \le K_0\phi_{i_0}(u_0) = K_0\frac{\phi_{i_0}(u_0)}{\phi_{i_0}(u')}\,\phi_{i_0}(u')$$

$$\le K'_0\phi_{i_0}(u') \le K'_0\phi_{i_0}(u).$$

Thus, we have proved that (5.2.22) is a necessary and sufficient condition for being essentially constant. To prove $L_\rho = L_{\rho_0}$, take $f \in L_\rho$. We have then that $\rho_i(\lambda f) \to 0$ as $\lambda \to 0$ for each index i separately. Since ϕ_i are equicontinuous at zero , for a fixed $\epsilon > 0$ there exists $u' > 0$ such that

$$\phi_i(u') < \tfrac{\epsilon}{2} \left(\sup_{\tau \in \mathcal{T}} \mu_\tau(X)\right)^{-1}$$

for every $i \in \mathbf{N}$. Let us pick an index i_0, and a positive number ω' from (5.2.22). We may, therefore, find $K'_0 > 0$ corresponding to u'. By (5.2.22) we obtain the following:

$$\phi_i(u) \leq K'_0 \phi_{i_0}(\omega' u) \text{ for } u \geq u' \text{ and } i \geq i_0.$$

Let $\lambda_0 > 0$ be so small that $\rho_i(\lambda f) < \epsilon$ for $i = 1, \ldots , i_0 - 1$ and

$$\rho_{i_0}(\omega' \lambda f) < \frac{\epsilon}{2K'_0}$$

for every $\lambda \in [0,\lambda_0]$. Define $A = \{t \in X; |\lambda f(t)| \geq u'\}$ and $B = X \setminus A$. For every $i \geq i_0$ and every $\lambda \in [0,\lambda_0]$ there holds

$$\rho_i(\lambda f) \leq \sup_{\tau \in \mathcal{T}} \int_A \phi_i(\lambda|f(t)|)d\mu_\tau + \sup_{\tau \in \mathcal{T}} \int_B \phi_i(u')d\mu_\tau$$

$$\leq K'_0 \sup_{\tau \in \mathcal{T}} \int_A \phi_{i_0}(\omega' \lambda|f(t)|)d\mu_\tau + \phi_i(u') \sup_{\tau \in \mathcal{T}} \mu_\tau(B) \leq \epsilon.$$

Finally, for all $i \in \mathbf{N}$ there holds $\rho_i(\lambda f) \leq \epsilon$ whenever $\lambda \leq \lambda_0$. Hence, $f \in L_{\rho_0}$.

(5.2.20) $\Rightarrow$ (5.2.21). Suppose to the contrary that the sequence (ϕ_k) is not essentially constant. For arbitrary indices n, m, k we can choose then an index i(n,m,k) $\geq$ n, a number u(n,m,k) $\geq$ m, such that

$$\phi_{i(n,m,k)}\left(2^{-k}u(n,m,k)\right) > 2^k \phi_n\left(u(n,m,k)\right).$$

Let us take a constant $\eta > 0$ which is mentioned in the definition of equisplittability of the family (μ_τ) and a sequence of positive numbers (v_i) from the definition of essential monotonicity of (ϕ_k). Choose $m_1 \geq v_1$ such that

$$\phi_1(m_1) \geq \min\left(1, \tfrac{1}{\eta}\right)$$

and put $u_1 = u(1,m_1,1)$. Next, define $m_2 \geq v_2$ so that

$$\phi_2(m_2) \geq \max\left(1, \phi_1(u_1)\right)$$

and denote $u_2 = u(2,m_2,2)$. Proceeding by induction we obtain two sequences (u_k) and (m_k) such that

$$\phi_k(u_k) \geq \phi_k(m_k) \geq \phi_{k-1}(u_{k-1}) > 0 \quad \text{for k} = 2, 3, \ldots .$$

Hence, the sequence $\left(\phi_k(u_k)\right)$ is increasing. Defining

$$\epsilon_k = \left(2^k \phi_k(u_k)\right)^{-1},$$

we observe that $\epsilon_k \downarrow 0$,

$$\epsilon_k \le \epsilon_1 \le \left(2\phi_1(u_1)\right)^{-1} \le \left(2\phi_1(m_1)\right)^{-1} \le \frac{\eta}{2} < \eta$$

and moreover

$$\frac{\epsilon_{k+1}}{\epsilon_k} = \frac{\phi_k(u_k)}{2\phi_{k+1}(u_{k+1})} < \frac{1}{2} \quad \text{for all } k \in \mathbb{N}.$$

Since (μ_τ) is equisplittable, it follows that there exist two constants $M \ge m > 0$ and a sequence (A_k) of pairwise disjoint sets $A_k \in \Sigma$ such that

$$(5.2.23) \qquad m\epsilon_k \le \sup_{\tau \in \mathcal{T}} \mu_\tau(A_k) \le M\epsilon_k \quad \text{for } k \in \mathbb{N}.$$

Let us define a measurable function f by the formula:

$$f(t) = \begin{cases} u_k & \text{for } t \in A_k, \\ 0 & \text{for } t \notin \bigcup_{i=1}^{\infty} A_i. \end{cases}$$

We will prove now that $f \in L_\rho = \bigcap_{n=1}^{\infty} L_{\rho_n}$. Indeed, by essential monotonicity, for every $n \in \mathbb{N}$ we have λ_n, β_n such that $\phi_n(\lambda_n u) \le \beta_n \phi_k(u)$ for $k \ge n$ and $u \ge v_i$. Thus,

$$\rho_n(\lambda_n f) = \sup_{\tau \in \mathcal{T}} \sum_{k=1}^{\infty} \phi_n(\lambda_n u_k)\, \mu_\tau(A_k)$$

$$\le \sup_{\tau \in \mathcal{T}} \sum_{k=1}^{n-1} \phi_n(\lambda_n u_k)\, \mu_\tau(A_k) + \sup_{\tau \in \mathcal{T}} \sum_{k=n}^{\infty} \phi_n(\lambda_n u_k)\, \mu_\tau(A_k)$$

$$\leq K \sum_{k=1}^{n-1} \phi_n(\lambda_n u_k) + \beta_n \sum_{k=n}^{\infty} \phi_k(u_k) \sup_{\tau \in \mathcal{J}} \mu_\tau(A_k),$$

where $K = \sup_{\tau \in \mathcal{J}} \mu_\tau(X) < \infty$. On the other hand, by (5.2.23)

$$\sum_{k=n}^{\infty} \phi_k(u_k) \sup_{\tau \in \mathcal{J}} \mu_\tau(A_k) \leq M \sum_{k=n}^{\infty} \phi_k(u_k) \, \epsilon_k = M \sum_{k=n}^{\infty} 2^{-k} = M \, 2^{1-n}$$

and consequently

$$\rho_n(\lambda_n f) \leq K \sup_{\tau \in \mathcal{J}} \sum_{k=n}^{\infty} \phi_n(\lambda_n u_k) + M \, \beta_n \, 2^{1-n} < \infty.$$

By Proposition 5.2.9 we conclude that $f \in L_{\rho_n}$. Since n was chosen arbitrarily, we obtain $f \in L_\rho$. On the other hand, the follwing computation proves that $\rho_n(\lambda f)$ does not tend to zero uniformly with respect to n as $\lambda \to 0$ and consequently $f \notin L_{\rho_0}$. Indeed,

$$\rho_{i(k,m_k,k)}(2^{-k}f) \geq \sup_{\tau \in \mathcal{J}} \int_{A_k} \phi_{i(k,m_k,k)}\left(2^{-k}|f(t)|\right) d\mu_\tau$$

$$= \phi_{i(k,m_k,k)}\left(2^{-k}u_k\right) \sup_{\tau \in \mathcal{J}} \mu_\tau(A_k)$$

$$\geq 2^k \, \phi_k(u_k) \sup_{\tau \in \mathcal{J}} \mu_\tau(A_k) \geq 2^k \, \phi_k(u_k) \, m\epsilon_k = m > 0.$$

Finally, $L_{\rho_0} \neq L_\rho$ which contradicts (a). This completes the proof in the case of modulars defined by (5.2.1); for the modulars given by (5.2.2) the result may be proved in the same manner.

5.2.24   REMARK.   Since the sequence $\phi_k(u) = |u|^{p_k}$ is a normal sequence of ϕ-functions provided (p_k) is nondecreasing and since ϕ_k is essentially constant if and only if (p_k) is constant for sufficiently large k, it follows that Theorem 5.2.19 is indeed a generalization of the result mention at the beginning of this chapter.

5.3 EXAMPLES OF KERNELS

Now we will present some examples of kernels for various methods of summation and will show that the measures defined by these kernels satisfy the assumptions of Theorem 5.2.19, i.e. are uniformly bounded and equisplittable.

5.3.1 PROPOSITION. Let $X = [t_0,\infty)$, $\mathcal{T} = [\tau_0,\infty)$, Σ be the σ-algebra of all Lebesgue measurable subsets of X and $a : X \times \mathcal{T} \to \mathbb{R}^+$ be measurable in the first variable. Suppose that there exist three numbers $\eta > 0$, $M \geq m > 0$ such that for every $\epsilon \in (0,\eta]$ and for every $s \geq t_0$ there exist $\beta > \alpha > s$ such that

$$m\epsilon \leq \sup_{\tau \in \mathcal{J}} \int_{\alpha}^{\beta} a(t,\tau)\, dt \leq M\epsilon.$$

Then, the family (μ_τ) is equisplittable, where

$$\mu_\tau(A) = \int_{A} a(t,\tau)\, dt; \quad A \in \Sigma.$$

PROOF.   Let $\epsilon_k \downarrow 0$, $\frac{\epsilon_{k+1}}{\epsilon_k} \leq \frac{1}{2}$. Put $s = t_0$, $\epsilon = \epsilon_1$, $\alpha_1 = \alpha$, $\beta_1 = \beta$ and $A_1 = (\alpha_1, \beta_1)$, where α and β exist by hypothesis. Next, put $s = \beta_1$, $\epsilon = \epsilon_2$, $\alpha_2 = \alpha$, $\beta_2 = \beta$, $A_2 = (\alpha_2, \beta_2)$. Proceeding by induction we obtain a sequence of disjoint sets (A_k) such that

$$m\epsilon_k \leq \sup_{\tau \in \mathcal{J}} \mu_\tau(A_k) = \sup_{\tau \in \mathcal{J}} \int_{\alpha_i}^{\beta_i} a(t,\tau)dt \leq M\epsilon_k$$

holds for all natural numbers k.

Similarly, we can immediately obtain the discrete version of this result.

5.3.2 PROPOSITION. Let $X = N$, and let Σ denote the σ-algebra of all subsets of X. Let $(a_{n,i})$ be an infinite matrix satisfying the condition of Definition 5.2.10. Recall that the measure μ_n was defined by

$$\mu_n(A) = \sum_{i \in A} a_{n,i} \quad \text{for A} \in \Sigma.$$

Suppose that there exist three numbers $\eta > 0$, $M \geq m > 0$ such that to every natural N there correspond two indices $q > p \geq N$ for which

$$m\epsilon \leq \sup_n \sum_{i=p}^{q} a_{ni} \leq M \epsilon.$$

Then, the family (μ_n) is equisplittable.

5.3.3 EXAMPLE. Cesàro kernel (Riesz means) of order $r \geq 1$:

$a(t,\tau) = \frac{r}{\tau}\left(1 - \frac{t}{\tau}\right)^{r-1}$ for $t \in [0,\tau]$,

$a(t,\tau) = 0$ for $t > \tau$,

$X = [0,\infty0$, $\mathcal{T} = (0,\infty)$.

We claim that the family of measures

$$\mu_T(A) = \int_A a(t,\tau)dt$$

is uniformly bounded and equisplittable. Indeed,

$$\mu_\tau(X) = \int_0^\infty a(t,\tau)dt = \int_0^\tau \frac{r}{\tau}\left(1-\frac{t}{\tau}\right)dt = 1,$$

i.e. (μ_τ) is uniformly bounded. In order to prove the equisplittability of (μ_τ) we will verify the assumptions of Proposition 5.3.1. Take $M = m = \frac{1}{2}$, fix $\epsilon > 0$ and $s > 0$ arbitrarily. For $\beta > \alpha \geq s$ there holds

$$\int_\alpha^\beta a(t,\tau)dt = \begin{cases} 0; & \tau \in (0,\alpha] \\ (1-\frac{\alpha}{\tau})^r; & \tau \in (\alpha,\beta) \\ (1-\frac{\alpha}{\tau})^r - (1-\frac{\beta}{\tau})^r; & \tau \in [\beta,\infty). \end{cases}$$

Hence, the above integral assumes its largest value at $\tau = \beta$ and

$$\sup_{\tau \in \mathcal{T}} \int_\alpha^\beta a(t,\tau)dt = \left(1 - \frac{\alpha}{\beta}\right)^r.$$

If we choose α and β so that $\frac{\alpha}{\beta} = 1 - \left(\frac{\epsilon}{2}\right)^{\frac{1}{r}}$, i.e. $\left(1 - \frac{\alpha}{\beta}\right)^r = \frac{\epsilon}{2}$, then

$$m\epsilon = \frac{\epsilon}{2} = \sup_{\tau \in \mathcal{T}} \int_\alpha^\beta a(t,\tau)dt = \left(1 - \frac{\alpha}{\beta}\right)^r = \frac{\epsilon}{2} = M\epsilon.$$

Thus, the family $(\mu_\tau)_{\tau \in \mathcal{T}}$ is equisplittable.

5.3.4 EXAMPLE. Stieltjes means of order $r > 0$:

$$a(t,\tau) = \frac{r}{\tau}\left(1 + \frac{t}{\tau}\right)^{-r-1}; \qquad X=[0,\infty), \quad \mathcal{T}=[t_0,\infty).$$

For every $\tau > 0$ there holds

$$\int_0^\infty a(t,\tau)\, dt = \int_0^\infty \frac{r}{\tau}\left(1 + \frac{t}{\tau}\right)^{-r-1} dt = 1,$$

i.e. the family of measures (μ_τ) is uniformly bounded. Since for every $\alpha > \beta \geq 0$ there holds

$$\int_\alpha^\beta a(t,\tau)\, dt = \left(1 + \frac{\alpha}{\tau}\right)^{-r} - \left(1 + \frac{\beta}{\tau}\right)^{-r},$$

it is reasonable to consider the function

$$h(\tau) = \left(1 + \frac{\alpha}{\tau}\right)^{-r} - \left(1 + \frac{\beta}{\tau}\right)^{-r}$$

which is a nonnegative, differentiable function, $\lim_{\tau \to 0} h(\tau) = 0$.

Moreover, $h'(\tau) = 0$ only if

$$\tau = \tau_0 = \beta\left(\left(\frac{\alpha}{\beta}\right)^{\frac{1}{1+r}} - \frac{\alpha}{\beta}\right)\left(1 - \left(\frac{\alpha}{\beta}\right)^{\frac{1}{1+r}}\right)^{-1}.$$

Hence,

$$\sup_\tau \int_\alpha^\beta a(t,\tau)\, dt = h(\tau_0) = \left(1 + \Theta\,\frac{1-\theta_r}{\theta_r-\theta}\right)^{-r} - \left(1 + \frac{1-\theta_r}{\theta_r-\theta}\right)^{-r},$$

where $\theta = \frac{\alpha}{\beta}$, $\theta_r = \theta^{\frac{1}{1+r}}$. The right-hand side of the above formula is a continuous function of θ; denote it by $g(\theta)$ and observe that $g(\theta) \to 0$ as $\theta \to 0^+$. It follows then that for every $\epsilon > 0$ we may find θ_0 so that $g(\theta_0) = \epsilon$. Hence, for every $s \geq 0$ we may choose $\beta > \alpha > s$ $\left(\beta = \frac{\alpha}{\theta_0}\right)$ such that

$$\sup_{\tau \in \mathcal{T}} \int_\alpha^\beta a(t,\tau)\,dt = \int_\alpha^\beta a(t,\tau_0)\,dt = \epsilon.$$

Hence, taking $M = m = 1$ we obtain by Proposition 5.3.1 that (μ_τ) are equisplittable.

5.3.5 EXAMPLE. Abel-Laplace means:

$$a(t,\tau) = \tau^{-1}e^{\frac{-t}{\tau}}; \quad X=[0,\infty], \quad \mathcal{T}=(0,\infty).$$

For every positive α and γ we may consider a function

$$h(\tau) = \int_\alpha^{\alpha+\gamma} \tau^{-1}e^{\frac{-t}{\tau}}dt = e^{\frac{-\alpha}{\tau}} - e^{\frac{-(\alpha+\gamma)}{\tau}} \geq 0;$$

h is a continuous function, $\lim_{\tau \to 0} h(\tau) = 0 = \lim_{\tau \to \infty} h(\tau)$. We observe, moreover, that $h'(\tau)=0$ if and only if

$$\tau=\tau_0 = \frac{\gamma}{\ln(1+\eta)}, \quad \text{where } \eta=\frac{\gamma}{\alpha}.$$

Thus, the function h achieves its maximum at τ_0. Let us define $\theta=\eta^{-1}$ and $g(\theta)= h(\tau_0) = \sup_{\tau \in \mathcal{T}} h(\tau)$. We have therefore,

$$g(\theta) = \sup_{\tau \in \mathcal{T}} \int_{\alpha}^{\alpha+\gamma} a(t,\tau)dt = e^{-\theta \ln(1+\eta)} - e^{-(\theta+1)\ln(1+\eta)} \to 0$$

as $\theta \to \infty$. Since the right-hand side of this formula is a continuous function of θ, it follows that to every $\epsilon > 0$ there exists θ_0 such that $g(\theta_0)=\epsilon$. Given $s > 0$, we may choose $\alpha_0 > 0$ and $\gamma_0 = \dfrac{\alpha_0}{\theta_0}$. Hence,

$$g(\theta_0) = \sup_{\tau \in \mathcal{T}} \int_{\alpha_0}^{\delta_0} a(t,\tau)dt = \epsilon, \quad \text{where } \delta_0 = \alpha_0 + \gamma_0.$$

The equisplittability of (μ_τ) follows via Proposition 5.3.1.

5.3.6 EXAMPLE. As examples of matrices (a_{n_i}) defining uniformly bounded and equisplitable families of measures (μ_n) we may take:

(a)   Césaro means (C,k) of orders $k \in \mathbb{N}$ defined by:

$$a_{n_i} = \begin{cases} \binom{n+k-i-1}{k-1}\binom{n+k}{k} & \text{for } i \leq n \\ \\ 0 & \text{for } i > n . \end{cases}$$

The proof of the fact that (μ_n) satisfies the hypothesis of

Proposition 5.3.2 may be found in Musielak [1].

(b)   Nörlund means (N, p_n) defined by:

$$a_{ni} = \begin{cases} \dfrac{P_{n-i}}{P_n} & \text{for } i \leq n \\[2mm] 0 & \text{for } i > n \end{cases}$$

where $p_n > 0$ are such that

$$P_n = p_0 + \cdots + p_n \rightarrow +\infty \quad \text{and} \quad \prod_{i=1}^{\infty} \frac{P_{i-1}}{P_i} < \infty.$$

Since for an arbitrary $q > p$,

$$\sup_n \sum_{i=p}^{q} a_{ni} = 1 - \frac{P_{p-1}}{P_q} \leq 1,$$

then the (μ_n) are uniformly bounded.

Fix $0 < \epsilon < 1$ and observe that the series

$$\sum_{i=1}^{\infty} \ln \frac{P_{i-1}}{P_i}$$

is convergent because

$$\prod_{i=1}^{\infty} \frac{P_{i-1}}{P_i} < \infty.$$

Then, for p, q sufficiently large:

$$-\ln \frac{P_{p\text{-}1}}{P_q} = \left| \sum_{i=p}^{q} \ln \frac{P_{i\text{-}1}}{P_i} \right| < -\ln(1-\epsilon)$$

which gives $1 - \dfrac{P_{p\text{-}1}}{P_q} < \epsilon$.

On the other hand, for fixed p and $q > p$ sufficiently large, $P_q > 2P_{p\text{-}1}$ (since $P_n \to +\infty$) and, therefore, we get:

$$\frac{P_{p\text{-}1}}{P_q} < \tfrac{1}{2} < 1 - \tfrac{1}{2}\epsilon.$$

Finally,

$$m\epsilon < 1 - \frac{P_{p\text{-}1}}{P_q} < M\epsilon,$$

where $M = 1$ and $m = \tfrac{1}{2}$, i.e. (μ_n) are equisplittable by Proposition 5.3.2.

We will prove now the topological version of Proposition 5.3.1.

5.3.7 PROPOSITION. Assume that there exists $M \geq m > 0$ such that to every $\epsilon \in (0,1]$ there corresponds $\vartheta \in (0,\epsilon]$ such that

$$m\epsilon \leq \limsup_{\tau \to \infty} \int_A a(t,\tau)dt \leq M\epsilon$$

where
$$A = \bigcup_{n=1}^{\infty} [\alpha_n, \beta_n)$$

and $[\alpha_n, \beta_n) \subset \mathbb{R}^+$ is an arbitrary sequence of disjoint intervals of length ϑ. Then, the family of measures (μ_n) defined as $\mu_\tau(A) = \int_A a(t,\tau) dt$ is topologically equisplittable.

PROOF. Given a sequence of numbers $\epsilon_k \downarrow 0$ such that

$$0 < \epsilon_k < 1 \text{ and } \frac{\epsilon_{k+1}}{\epsilon_k} \leq \frac{1}{2};$$

put $\alpha_n^1 = n-1$, $\beta_n^1 = (1-\vartheta_1)(n-1) + \vartheta n$, where ϑ_1 corresponds to $\epsilon = \epsilon_1$. By the assumption, the set $A_1 = \bigcup_{n=1}^{\infty} [\alpha_n^1, \beta_n^1)$ satisfies

$$m\epsilon \leq \limsup_{\tau \to \infty} \int_{A_1} a(t,\tau) dt \leq M\epsilon.$$

Proceeding by the induction we get a sequence $\alpha_n^{k+1} = \beta_n^k$, $\beta_n^k - \alpha_n^k = \vartheta_k$, $\beta_n^{k+1} = \alpha_n^{k+1} + \vartheta_{k+1}$, where $\vartheta_k \in (0, \epsilon_k]$;   we may observe that $\bigcup_{k=1}^{\infty} [\alpha_n^k, \beta_n^k) \subset [n-1, n]$ because

$$\sum_{k=1}^{\infty} (\beta_n^k - \alpha_n^k) \leq \sum_{k=1}^{\infty} \epsilon_k < 1.$$

Hence, defining $A_m = \bigcup_{n=1}^{\infty} [\alpha_n^m, \beta_n^m)$, we can obtain a sequence of disjoint sets such that

$$m\epsilon_k \leq \limsup_{\tau \to \infty} \int_{A_k} a(t,\tau) dt \leq M\epsilon_k.$$

5.3.8 EXAMPLE. The following families of measures are topologically equisplittable (for definitions see Examples 5.3.3, 5.3.4, 5.3.5).

(a)   Cesaro kernel of order $r \geq 1$: Put $m = 4^{-r}$, $M = 1$. Given ϵ, define $\vartheta = \frac{\epsilon}{r}$. Suppose $\tau \in [\beta_{n-1}, \alpha_n]$; then

$$\int_A a(t,\tau)dt = \sum_{i=1}^{n-1} \left\{ \left(1 - \frac{\alpha_i}{\tau}\right)^r - \left(1 - \frac{\beta_i}{\tau}\right)^r \right\}$$

$$\leq \frac{\vartheta}{\tau} r(n-1) \leq \epsilon \frac{\tau+1}{\tau} \to \epsilon \quad \text{as } \tau \to \infty.$$

On the other hand,

$$\int_A a(t,\tau)dt \geq \frac{\vartheta}{\tau} r \sum_{i=1}^{E(n)} \left(1 - \frac{\gamma_i}{\tau}\right)^{r-1}$$

$$\geq \frac{\epsilon}{\tau} E(n) 4^{1-r} \geq \epsilon \frac{\tau-3}{\tau} \cdot 4^{-r} \geq m\epsilon,$$

where $\gamma_i \in [\alpha_i, \beta_i]$ and $E(n)$ is the integer part of the number $\frac{1}{2}(n-1)$. In the other case $\left(\tau \in (\alpha_n, \beta_n)\right)$ we may similarly obtain

$$m\epsilon \leq \int_A a(t,\tau)dt < \epsilon + \tau^{-k} \to \epsilon \quad \text{as } \tau \to \infty.$$

(b)   Stieltjes kernel of order $\rho > 0$:

Let $M = m = 1$. Given $\epsilon > 0$, put $\vartheta = \epsilon$. We have

$$\int_A a(t,\tau)dt = \frac{\rho\vartheta}{\tau} \sum_{i=1}^{\infty} \left(1 + \frac{\gamma_n}{\tau}\right)^{-\rho-1} \le \frac{\rho\epsilon}{\tau}\left(1 + \frac{\tau}{\rho}\right) \to \epsilon$$

as $\tau \to \infty$, where $\gamma_n \in [\alpha_n, \beta_n]$. On the other hand,

$$\int_A a(t,\tau)dt \ge \frac{\rho\epsilon}{\tau}\frac{\tau}{\rho}\left(1 + \frac{1}{\tau}\right)^{-\rho} \to \epsilon \text{ as } \tau \to \infty.$$

(c)   Abel-Laplace means:

Using the same notation as in the previous example we may calculate

$$\epsilon e^{\frac{-1}{\tau}} \le \int_A a(t,\tau)dt \le \epsilon \frac{1 + \tau}{\tau}.$$

5.3.9 EXAMPLE. The family of the Dirac measures $(\delta_\tau)_{\tau \ge \tau_0}$ is not equisplittable, because we have   for every nonempty measurable set $A \subset \mathbb{R}^+$: $\sup_{\tau \ge \tau_0} \delta_\tau(A) = 1.$

5.3.10 EXAMPLE. Given a sequence (ϕ_k) of normal ϕ-functions and $\epsilon > 0$. One may ask the following question:

When is the following infinite system of inequalities

$$\limsup_{\tau \to \infty} \frac{1}{\tau} \int_0^{\tau} a(t,\tau) \, \phi_1\Big(\lambda f(t)\Big) dt < \epsilon$$

for $\lambda > 0$ sufficiently small,

$$\vdots$$

$$\lim_{\tau \to \infty} \sup \frac{1}{\tau} \int_0^\tau a(t,\tau)\ \phi_k\Big(\lambda f(t)\Big)dt < \epsilon$$

for $\lambda > 0$ sufficiently small,

$$\vdots$$

equivalent (f being an unknown measurable function) to only one inequality

$$\sup_{k \in \mathbb{N}} \lim_{\tau \to \infty} \sup \frac{1}{\tau} \int_0^\tau a(t,\tau)\ \phi_k\Big(\lambda(t)\Big)dt < \epsilon$$

for $\lambda > 0$ sufficiently small ?

From Theorem 5.2.19 it can be easily deduced that this equivalence may hold only if (ϕ_k) is an essentially constant sequence of ϕ-functions.

5.3.11 REMARK. Theorem 5.2.19 will be applied in Chapter 6 in the discussion of a theory of nonlinear operators defined in modular function spaces.

Bibliographical remarks

The idea of countably modulared and uniformly countably modulared spaces was introduced by J. Albrycht and J. Musielak [1] in 1968. Kozlowski considered in [5] these same constructions for function modulars. The results related to the theory of summation and various kernels were elaborated in 1967-1971 by J. Musielak and A. Waszak [1], [2], [3], [4] and by A. Waszak in [1] and [2]. The notion of an equisplittable and topologically equisplittable family of measures as well as other related problems are discussed in the book of Musielak [1].

6. Nonlinear Operators

In Section 6.1 we study the problem of extending the domain of a nonlinear, disjointly additive operator from a space $\mathcal{E}$ of all $\mathcal{P}$-simple functions to a larger F-space of measurable functions. Theorem 6.1.9 is the main result of this section. In Section 6.2 we give some examples and determine the extended domains for Nemytskii, Hammerstein and nonlinear Fourier operators. The next section deals with a problem of equicontinuity of a sequence of nonlinear operators while in Section 6.4 we present some results on fixed points of operators that are nonexpansive in a modular sense.

6.1 EXTENSION OF NONLINEAR OPERATORS

Suppose that T is a disjointly additive operator which maps the space $\mathcal{E}$ of all $\mathcal{P}$-simple functions into an F-space $(H, \|\cdot\|_H)$. We want to extend the domain of definition of T to a larger subset of $M(X,S)$ and construct a function F-space (see the definition below) $E \subset M(X,S)$ such that $T : E \to H$ is a continuous operator. We will also study a maximal property of our construction.

6.1.1 DEFINITION. An F-space $(E, \|\cdot\|_E)$, $E \subset M(X,S)$ will be called a function F-space if

$$E = \{ f \in M(X,S) ; \|\lambda f\|_E \to 0 \text{ as } \lambda \to 0 \}.$$

and $\|f\|_E \leq \|g\|_E$ provided $|f(x)| \leq |g(x)|$, $x \in X$.

6.1.2. REMARK. For a nonlinear operator $T : \mathcal{E} \to H$ where H is an arbitrary F- space, for every $\alpha > 0$ and $A \in \mathcal{P}$ we define

$$\omega_\delta(T,\alpha,A)$$

$$= \sup\{\|T(f) - T(g)\|_H; f, g \in \mathcal{E}, \text{supp}(f) \subset A, \text{supp}(g) \subset A,$$

$$|f(x)| \leq \alpha, |g(x)| \leq \alpha, |f(x) - g(x)| \leq \delta \text{ for all } x \in X \}.$$

Let us define the following (α,A)-majorant of T, where $\alpha > 0$, $A \in \mathcal{P}$:

$$\overline{T}(\alpha,A) = \sup\{\|T(g)\|; g \in \mathcal{E}, |g| \leq \alpha 1_A\}.$$

Throughout this section we will assume that the operator $T : \mathcal{E} \to H$ satisfies

(6.1.3) T is disjointly additive, i.e. $T(f + g) = T(f) + T(g)$ for every f, g $\in \mathcal{E}$ with disjoint supports,

(6.1.4) For every $\alpha > 0$, $A \in \mathcal{P}$, $\omega_\delta(T,\alpha,A) \to 0$ as $\delta \to 0$,

(6.1.5) For every $\alpha > 0$ and $E_k \downarrow \emptyset$, $\overline{T}(\alpha,E_k) \to 0$ as $k \to \infty$,

(6.1.6) There exists a number $\alpha_0 > 0$ such that $\overline{T}(\beta,A) = 0$ for every $\beta > 0$ whenever $\overline{T}(\alpha,A) = 0$ for a certain $\alpha > \alpha_0$, A being an arbitrarily fixed set from $\mathcal{P}$.

6.1.7 DEFINITION. Let us define the functional ρ by

$$\rho(f,A) = \sup\{\|T(g)\|_H; g \in \mathcal{E}, |g(x)| \leq 1_A(x)|f(x)| \text{ for } x \in X\},$$

where $f \in \mathcal{E}$, $A \in \Sigma$.

From the above definition and the conditions (6.3.1), (6.3.2), (6.3.3), (6.3.4) we get the following result:

6.1.8 PROPOSITION. The functional $\rho : \mathcal{E} \times \Sigma \to [0,\infty]$ is a function semimodular.

PROOF. Since T is additive then $T(0) = 0$ and therefore (P1) holds. The monotonicity of ρ is an immediate consequence of the following inclusion which holds for $A \in \Sigma$, f, g $\in \mathcal{E}$, $|f(x)| \leq |g(x)|$ for all $x \in A$:

$$\{\|T(h)\|_H \; ; \; h \in \mathcal{S}, \; |h(x)| \leq 1_A(x) \; |f(x)| \text{ for } x \in X\}$$

$$\subset \{\|T(h)\|_H \; ; \; h \in \mathcal{S}, \; |h(x)| \leq 1_A(x) \; |g(x)| \text{ for } x \in X\}.$$

In order to prove (P3) let us observe first that $\rho(f,\emptyset) = 0$ since $1_\emptyset = 0$. Moreover, if $A \subset B$ then $\rho(f,A) \leq \rho(f,B)$ because

$$\{\|T(h)\|_H \; ; \; h \in \mathcal{S}, \; |h(x)| \leq 1_A(x) \; |f(x)| \text{ for all } x \in X\}$$

$$\subset \{\|T(h)\|_H \; ; \; h \in \mathcal{S}, \; |h(x)| \leq 1_B(x) \; |f(x)| \text{ for all } x \in X\}.$$

To prove the σ-subadditivity of $\rho(f,\cdot)$ let us take first a sequence (A_n) of mutually disjoint sets from Σ and let us denote $A = \bigcup_{n=1}^{\infty} A_n$. Let us take a function $g \in \mathcal{S}$ such that $\mathrm{supp}(g) \subset A$ and compute

$$\|T(g)\|_H = \|T(g1_A + T(g1_{A'})\|_H \leq \|T(g1_A)\|_H$$

$$\leq \sum_{i=1}^{N} \|T(g1_{A_n})\|_H + \|T(g1_{B_N})\|_H,$$

where $A' = X \setminus A$ and $B_N = \bigcup_{n=N+1}^{\infty} A_n$. Since $g \in \mathcal{S}$ and $B_N \downarrow \emptyset$ it follows from (6.1.5) that

$$\|T(g1_{B_N})\|_H \rightarrow 0 \quad \text{as} \quad N \rightarrow \infty.$$

Hence

$$\|T(g)\| \leq \sum_{i=1}^{\infty} \|T(g1_{A_n})\|_H.$$

Compute

$$\rho(f, A) = \sup\{\|T(g)\| \; ; \; g \in \mathcal{E}, |g(x)| \leq 1_A(x) |f(x)|, x \in X\}$$

$$\leq \sup\left\{ \sum_{n=1}^{\infty} \|T(g1_{A_n})\|_H; \; g \in \mathcal{E}, |g(x)| \leq 1_A(x) |f(x)|, x \in X\right\}$$

$$\leq \sum_{n=1}^{\infty} \sup\{\|T(g1_{A_n})\|_H; \; g \in \mathcal{E}, |g(x)| \leq 1_A(x) |f(x)|, \; x \in X\}$$

$$\leq \sum_{n=1}^{\infty} \sup\{\|T(g1_{A_n})\|_H; \; g \in \mathcal{E}, 1_{A_n}(x) |g(x)| \leq 1_{A_n}(x) |f(x)|\}$$

$$\leq \sum_{n=1}^{\infty} \sup\{\|T(h)\|_H; \; h \in \mathcal{E}, |h(x)| \leq 1_{A_n}(x) |f(x)|, x \in X\}$$

$$\leq \sum_{n=1}^{\infty} \rho(f, A_n).$$

Finally,

$$\rho\left(f, \bigcup_{n=1}^{\infty} A_n\right) \leq \sum_{n=1}^{\infty} \rho(f, A_n)$$

for disjointly (A_n). Since

$$\bigcup_{n=1}^{\infty} A_n = \bigcup_{n=1}^{\infty} C_n, \text{ where } C_n = A_n \setminus \bigcup_{i=1}^{n-1} A_i,$$

then

$$\rho\left(f, \bigcup_{n=1}^{\infty} A_n\right) = \rho\left(f, \bigcup_{n=1}^{\infty} C_n\right) \leq \sum_{n=1}^{\infty} \rho(f, C_n) \leq \sum_{n=1}^{\infty} \rho(f, A_n).$$

The property (P4) follows from (6.1.4) via

$$\rho(\alpha, A) \leq \omega_\alpha(T, \alpha_0, A) \rightarrow 0 \text{ as } \alpha \rightarrow 0.$$

The property (P5) follows immediately from (6.1.6) while (P6) has been assumed in (6.1.5).

6.1.9 THEOREM.

(i)   We can extend T to the set L_ρ^0 by the formula
$$T(f) = \lim_{n \rightarrow \infty} T(f_n), \text{ where } f_n \in \mathcal{S}, |f_n(x)| \leq |f(x)|,$$
$f_n(x) \rightarrow f(x)$ for all $x \in X$.

(ii)   This definition does not depend on the choice of such a sequence of simple functions (f_n).

(iii)   The operator $T : L_\rho^0 \rightarrow H$ is additive.

(iv)   The inequality $\|T(f1_A)\|_H \leq \rho(f, A)$ holds for every $f \in L_\rho^0$, $A \in \Sigma$.

(v) $\|T(f1_{A_k})\|_H \rightarrow 0$ whenever $f \in L_\rho^0$, $A_k \downarrow \emptyset$.

(vi) $T : E_\rho \rightarrow H$ is a continuous operator.

(vii) E_ρ is the largest function F-space with the Lebesgue property (i.e., with absolutely continuous F-norm) such that T is continuous in E_ρ.

PROOF.

(i). Take a function $f \in L_\rho^0$; since f is measurable, it follows that there exists a sequence of simple functions (f_n) such that $|f_n(x)| \leq |f(x)|$ and $f_n(x) \rightarrow f(x)$ for all $x \in X$. By Egoroff's

Theorem we can pick up a sequence of sets $H_k \in \mathcal{P}$ such that $\overset{\infty}{\underset{n=1}{\bigcup}} H_k = X$ and (f_n) converges uniformly to the function f on every set H_k. Let us fix arbitrarily $\epsilon > 0$. Since $f \in L^0_\rho$, we may find such natural k that $\rho(f, D_k) < \frac{\epsilon}{4}$ where $D_k = X \setminus H_k$. We may find $G_k \subset H_k$ so that the function $f1_{G_k}$ is bounded and $\rho(f, H_k \setminus G_k) < \frac{\epsilon}{8}$. Consequently, the functions $f_n 1_{G_k}$, $f_m 1_{G_k}$ are bounded. Denote $Z_k = H_k \setminus G_k$ and compute

$$\|T(f_n 1_{H_k}) - T(f_m 1_{H_k})\|$$

$$\leq \|T(f_n 1_{G_k}) - T(f_m 1_{G_k})\| + \|T(f_n 1_{Z_k})\| + \|T(f_m 1_{Z_k})\|$$

$$\leq \omega_{\delta n,m}(T, \alpha, G_k) + 2\rho(f, Z_k) \leq \frac{\epsilon}{2}$$

for n, m sufficiently large, where $\alpha > 0$ was chosen such that

$$\sup\{|f_n(x)|;\ x \in G_k\} < \alpha \quad \text{for } n \geq N_0$$

and

$$\delta_{n,m} = \sup\{|f_n(x) - f_m(x)|;\ x \in G_k\} \to 0 \text{ as } n, m \to \infty.$$

We have

$$\|T(f_n) - T(f_m)\|$$

$$\leq \|T(f_n 1_{H_k}) - T(f_m 1_{H_k})\| + \|T(f_n 1_{D_k})\| + \|T(f_m 1_{D_k})\|$$

$$\leq \|T(f_n 1_{H_k}) - T(f_m 1_{H_k})\| + 2\rho(f, D_k) < \epsilon.$$

We conclude that the limit $\lim\limits_{n \to \infty} T(f_n)$ exists because H is complete.

(ii). The limit does not depend on the choice of the sequence (f_n). Indeed, take another $g_n \in \mathcal{S}$, $|g_n(x)| \leq |f(x)|$ and $g_n(x) \to f(x)$ for all $x \in X$; we may take then the sequence $(f_1, g_1, f_2, g_2, ...)$ which satisfies the same conditions. Consequently, $\lim\limits_{n \to \infty} T(f_n) = \lim\limits_{n \to \infty} T(g_n)$.

(iii).   This is evident in view of additivity of T on simple functions.

(iv).   Given $f \in L_\rho^0$ and $A \in \Sigma$. We have $T(f1_A) = \lim\limits_{n \to \infty} T(s_n 1_A)$ where $s_n \in \mathcal{S}$, $|s_n| \uparrow |f|$. Fix arbitrary $\epsilon > 0$. For n sufficiently large there holds

$$\|T(f1_A) - T(s_n 1_A)\| < \epsilon.$$

Hence,

$$\|T(f1_A)\| < \epsilon + \|T(s_n 1_A)\| \leq \epsilon + \rho(f, A),$$

which yields

$$\|T(f1_A)\| \leq \rho(f, A).$$

(v).   Immediate consequence of (iv).

(vi).   In order to get continuity of $T : E_\rho \to H$ we will prove that every sequence (f_n) converging to f in E_ρ contains a subsequence (g_n) such that $\|T(g_n) - T(f)\| \to 0$. Assume, therefore, that the functions f_n, $f \in E_\rho$ and $\|f_n - f\|_\rho \to 0$. There exists a subsequence (g_n) of (f_n) such that $g_n \to f$ ρ-a.e. By Egoroff's Theorem we can select a sequence (H_k) such that $H_k \in \mathcal{P}$, $H_k \uparrow X$ and (g_n) converges uniformly to f on every H_k. For given $\epsilon > 0$ let us take $k \in \mathbb{N}$ such that $\rho(2f,W_k) < \frac{\epsilon}{8}$, where $W_k = X \setminus H_k$ (recall $f \in E_\rho$) and $n_1 \in \mathbb{N}$ such that $\rho(2(g_n - f)) < \frac{\epsilon}{8}$ for $n \geq n_1$. Thus,

$$\rho(g_n,W_k) \leq \rho(2(g_n - f),W_k) + \rho(2f,W_k) < \frac{\epsilon}{4}.$$

Let us consider the following inequalities $(n \geq n_1)$:

$$\|T(g_n)-T(f)\|$$

$$\leq \|T(g_n 1_{H_k}) - T(f 1_{H_k})\| + \|T(g_n 1_{W_k})\| + \|T(f 1_{W_k})\|$$

$$\leq \|T(g_n 1_{H_k}) - T(f 1_{H_k})\| + \rho(g_n,W_k) + \rho(f,W_k)$$

$$< \|T(g_n 1_{H_k}) - T(f 1_{H_k})\| + \frac{\epsilon}{4} + \frac{\epsilon}{8}$$

$$\leq \|T(g_n 1_{H_k})-T(f 1_{H_k})\| + \frac{\epsilon}{2}.$$

Let (G_m) be a sequence of sets from $\mathcal{P}$ such that $G_m \uparrow H_k$ and $f 1_{G_m}$ are bounded for all m. Similarly as was done above we can take m and $n_2 \geq n_1$ such that

$$\rho(2f, H_k \setminus G_m) < \tfrac{\epsilon}{4} \text{ for } n \geq n_2.$$

Since (g_n) converges uniformly to f on G_m, it follows by (6.1.4) that $\|T(g_n 1_{G_m}) - T(f 1_{G_m})\| < \epsilon$ for $n \geq n_3 \geq n_2 \geq n_1$. Hence,

$$\|T(g_n 1_{H_k}) - T(f 1_{H_k})\|$$

$$< \epsilon + \|T(g_n 1_{H_{k,m}})\| + \|T(f 1_{H_{k,m}})\| \leq \epsilon + \tfrac{\epsilon}{2},$$

where $H_{k,m} = H_k \setminus G_m$. Finally, $\|T(g_n) - T(f)\| < 2\epsilon$, which completes the proof of (vi).

(vii). Assume $E \subset M$ is a function F-space with the Lebesgue property such that $T : E \rightarrow H$ is continuous but $E \setminus E_\rho \neq \emptyset$. Take a function $f \in E \setminus E_\rho$; there exists a sequence $D_k \downarrow \emptyset$ such that $\|f 1_{D_k}\|_\rho$ does not converge to zero, which gives (passing to a subsequence if necessary) positive constants γ, α such that

$$\rho(\alpha f 1_{D_k}) > \gamma > 0 \text{ for all natural k.}$$

There exist, therefore, functions $g_k \in \mathcal{E}$ such that $|g_k| \leq \alpha |f 1_{D_k}|$ and $\|T(g_k)\|_H > \tfrac{\gamma}{2}$. By the monotonicity and the absolute continuity of the F-norm in E we may conclude however that

$$\|g_k\|_E \leq \|\alpha f 1_{D_k}\|_E \to 0.$$

In view of the continuity of $T : E \to H$ we get

$$0 < \frac{\gamma}{2} \leq \|T(g_k)\|_E \to 0.$$

The contradiction completes the proof of (vii) and of the entire Theorem 6.1.9.

Let us assume again that the operator T acts from $\mathcal{E}$ into H, where H is a function F-space of measurable functions $f : X \to S$. Assume also that H has the Lebesgue property, i.e. $\|f 1_{E_k}\|_H \to 0$ as $E_k \downarrow \emptyset$. Let us observe that the F-norm $\|\cdot\|_H$ is also a function modular and therefore its Lebesgue property means simply that ρ satisfies the Δ_2-condition. Suppose $T : \mathcal{E} \to H$ satisfies the following condition:

(1)   T is additive, i.e. T satisfies (6.1.3);

(2)   T is Lipschitzian with constant L, i.e.
$$\|T(f) - T(g)\|_H \leq L \|f - g\|_H \text{ for all } f, g \in H;$$

(3)   The condition (6.1.6) holds.

Let us take $\alpha > 0$, $\delta > 0$, $A \in \mathcal{P}$. We have

$$\omega_\delta(T,\alpha,A)$$

$$\leq L \sup\{\|f - g\|_H \ ; \ f, g \in \mathcal{E}, \ supp(f) \subset A, \ supp(g) \subset A,$$

$$|f(x)| \leq \alpha, \ |g(X)| \leq \alpha, \ |f(x) - g(x)| \leq \delta\}$$

$$\leq L \ \|r1_A\|,$$

where $r \in S$ and $|r| = \delta$. Consequently,

$$\omega_\delta(T,\alpha,A) \rightarrow 0 \text{ as } \delta \rightarrow 0.$$

Similarly, if $\alpha > 0$ and $E_k \downarrow \emptyset$, then

$$\overline{T}(f,E_k) = \sup\{\|T(f)\|_H \ ; \ f \in \mathcal{E}, \ |f(x)| \leq \alpha 1_{E_k}(x) \ \rho\text{-a.e.}\}$$

$$\leq \sup\{\|T(f) - T(0)\|_H \ ; \ f \in \mathcal{E}, \ |f(x)| \leq \alpha 1_{E_k}(x) \ \rho\text{-a.e.}\}$$

$$\leq L \sup\{\|f\|_H \ ; \ f \in \mathcal{E}, \ |f(x)| \leq \alpha 1_{E_k}(x) \ \rho\text{-a.e.}\}$$

$$\leq L\alpha\|1_{E_k}\|_H \rightarrow 0.$$

We have just showed that T satisfies conditions (6.1.3) through (6.1.6). In view of Theorem 6.1.9, hence we can construct a function semimodular ρ_T induced by T. Let us introduce another function semimodular by the formula

$$\rho_H(f,A) = \rho_T(f,A) + \|f1_A\|_H.$$

6.1.10 PROPOSITION. $E_{\rho_H} = H$.

PROOF. Indeed, let $f \in H$, $E_k \downarrow \emptyset$. Take any $\epsilon > 0$ and $g \in \mathcal{S}$ such that $|f(x)| \le |g(x)|$ for all $x \in X$. We have

$$\|T(g1_{E_k})\|_H \le \|g1_{E_k}\|_H \le \|f1_{E_k}\|_H.$$

Since $\|\cdot\|_H$ is absolutely continuous, it follows that for k sufficiently large $\|f1_{E_k}\|_H < \frac{\epsilon}{2}$. Hence,

$$\rho_H(f1_{E_k}) = \rho_T(f1_{E_k}) + \|f1_{E_k}\|_H$$

$$= \sup\{\|T(g1_{E_k})\|_H \; ; g \in \mathcal{S}, |g(x)| \le 1_{E_k}(x)|f(x)|\} + \|f1_{E_k}\|_H$$

$$\le 2 \|f1_{E_k}\|_H < \epsilon$$

for k sufficiently large. This states that $f \in E_{\rho_H}$. Conversely, if $f \in E_{\rho_H}$ then $\rho_H(\lambda f) \to 0$ as $\lambda \to 0$; that implies $\|\lambda f\|_H \to 0$, i.e. f belongs to H.

6.1.11 REMARK. Note that $T : (H, \rho_H) \rightarrow (H, \|\cdot\|_H)$ is continuous, $\mathcal{E}$ is dense in (H, ρ_H), T satisfies the Lipschitz condition in $\mathcal{E}$ and therefore T satisfies the Lipschitz condition (with the same constant) in H.

6.2 EXAMPLES

6.2.1 EXAMPLE. Let us consider the Nemytskii operator $T(f)(x) = \phi(x, f(x))$ defined for every simple, Lebesgue measurable function $f : I \rightarrow \mathbb{R}$ where $I = [0,1]$, ϕ is a ϕ-function and $H = L^1(I, m)$; m denotes here the Lebesgue measure in I. Assume additionally that the function $\phi : I \times \mathbb{R} \rightarrow \mathbb{R}^+$ is continuous, the function $\phi(x, \cdot)$ is even for every $x \in I$ and the function $\phi(\cdot, \alpha)$ is summable for every positive α. Then the condition (6.1.4) follows easily by the uniform continuity of ϕ on every set of the form $I \times [0, \alpha]$, while the summability of the function $\phi(\cdot, \alpha)$ gives immediately (6.1.5). The property (6.1.6) follows from the fact that $\phi(x, u) > 0$ for $u > 0$. Take any measurable $f : I \rightarrow \mathbb{R}$. We have

$$\rho(f) = \sup\left\{ \int_I \phi\big(x, g(x)\big) \, dm(x); \, |g| \leq |f|, \, g \in \mathcal{E} \right\}$$

$$= \int_I \phi\big(x, f(x)\big) \, dm(x),$$

which means simply that ρ is the Musielak-Orlicz modular. In view of Theorem 6.1.9 the Nemytskii operator T acts continuously from E_ρ to L^1. In this case, E_ρ coincides with E^ϕ, the subspace of all finite elements of the Orlicz space L^ϕ, defined by

$$E^\phi = \left\{ f \in M; \int_I \phi\big(x, \lambda f(x)\big) \, dm(x) < \infty \quad \text{for every } \lambda > 0 \right\}.$$

6.2.2 EXAMPLE . Let ϕ be a ϕ-function which satisfies the following Lipschitz - type condition:

to every $\alpha > 0$ there exists a constant M_α such that

$$(6.2.2.\text{a}) \qquad |\phi(x,u) - \phi(x,v)| \le M_\alpha |u - v|$$

holds for every u, v with $|u|$, $|v| \le \alpha$ and for almost all x in $\mathbb{R}$.

Let $S = \mathbb{R}$, and $M = M(X,\mathbb{R})$. Assume again that the function $\phi(\cdot,\alpha)$ is summable for every $\alpha > 0$ and that $\phi(x,\cdot)$ is even for every x in X. Let us assume that $k : \mathbb{R} \times \mathbb{R} \to \mathbb{R}_+$ is a measurable function. By Σ we denote the σ-algebra of Lebesgue measurable subsets from $\mathbb{R}$, and $\mathcal{P}$ stands for the δ-ring of sets of finite measure. Assume that $k(x,y) > 0$ for all $x \in \mathbb{R}$ and $y \in Z$, where $Z \subset \mathbb{R}$ is of positive measure. Let us consider the Hammerstein operator

$$T(f)(x) = \int_\mathbb{R} k(x,y) \, \phi\big(y, f(y)\big) \, dm(y), \quad f \in \mathcal{S}.$$

The operator T is additive ; (6.1.4) follows from (6.2.2.a). If $H = L^\infty(\mathbb{R})$, then

$$\rho(f) = \operatorname*{ess\,sup}_{x\in\mathbb{R}} \int_{\mathbb{R}} k(x,y)\ \phi\Big(y,f(y)\Big)\ dm(y)$$

and E_ρ consists of all $f \in M$ such that for every $\lambda > 0$ and every $D_k \downarrow \emptyset$

$$\operatorname*{ess\,sup}_{x\in\mathbb{R}} \int_{D_k} k(x,y)\ \phi\Big(y,\lambda|f(y)|\Big)\ dm(y) \to 0\ .$$

Suppose that the kernel is degenerate, i.e. the function k is bounded and does not depend on the second variable; then (6.1.5) easily follows from the summability of functions of the form $\phi(\cdot,\alpha)$. (6.1.6) holds in view of strict positivity of ϕ on $\mathbb{R} \times Z$. Moreover, $E_\rho = E^\phi$. Assume now that for a $p > 1$ there exists a constant M such that

$$\int_{\mathbb{R}} k(x,y)^P\ dm(y) < M < \infty\ \text{ holds for all } x \in \mathbb{R}.$$

In view of the Hölder inequality, we get then that the condition (6.1.5) is satisfied and that $E_\rho \supset E^\psi$, where $\psi(x,u) = [\phi(x,u)]^q$, $p^{-1}+q^{-1} = 1$. Indeed,

$$\operatorname*{ess\,sup}_{x\in\mathbb{R}} \int_{B} k(x,y)\ \phi\Big(y,f(y)\Big)\ dm(y)$$

$$\operatorname*{ess\,sup}_{x \in \mathbb{R}} \left(\int_B [k(x,y)]^p \, dm(y) \right)^{\frac{1}{p}} \left(\int_B [\phi(y,f(y))]^q \, dm(y) \right)^{\frac{1}{q}}$$

$$\leq M^{\frac{1}{p}} \left(\int_{A_k} [\phi(y,f(y))]^q \, dm(y) \right)^{\frac{1}{q}} \to 0.$$

The strict inclusion may hold here when the kernel is not degenerate. For instance, let $k(x,y)=0$ for $y \geq 0$ and all $x \in \mathbb{R}$ while $k(x,y) > 0$ otherwise. If $D \subset \mathbb{R}^+$, $\lambda > 0$ then

$$\operatorname*{ess\,sup}_{x \in \mathbb{R}} \int_D k(x,y) \, \phi\big(y,\lambda|f(y)|\big) \, dm(y) = 0$$

holds for every measurable function f. Hence, for every sequence of measurable sets $A_k \downarrow \emptyset$ there holds

$$\operatorname*{ess\,sup}_{x \in \mathbb{R}} \int_{A_k} k(x,y) \, \phi\big(y,\lambda|f(y)|\big) \, dm(y) \to 0$$

for every function f such that $f1_{\mathbb{R}^-}$ belongs to E^ψ, even when f itself is not a member of E^ψ. Let us also observe that the space E_ρ will be different from any Musielak-Orlicz space if we drop the positivity or monotonicity of the function ϕ.

6.2.3 **EXAMPLE.** Consider the operator

$$T(f)(x) = \int_{\mathbb{R}} e^{-itx} \, \phi\big(f(t)\big) \, dm(t),$$

i.e. the nonlinear version of the Fourier transformation. The function $\phi : \mathbb{R} \to \mathbb{R}$ is an increasing, continuous, odd function such that

$$\lim_{u \to +\infty} \phi(u) = +\infty \text{ and } \lim_{u \to -\infty} \phi(u) = -\infty.$$

We will prove that $E_\rho = E^\psi$, where ρ is a function modular induced by the nonlinear operator T and $\psi(u) = [\phi(u)]^2$. Let $f \in E^\psi$. Since for every constant $\alpha > 0$ and every g in $\mathcal{S}$ the function $\phi(\alpha g(\cdot))$ belongs to L^2 and the Fourier transformation $\mathcal{F} : L^2 \to L^2$ is an isometry then we get the following sequence of equalities:

$$\rho(\alpha f, A_k) =$$

$$= \sup\left\{ \left(\int_{\mathbb{R}} \left| \int_{A_k} e^{-itx} \phi\big(\alpha g(t)\big) \, dm(t) \right|^2 dm(x) \right)^{\frac{1}{2}} ; g \in \mathcal{S}, |g| \le |f| \right\}$$

$$= \sup\left\{ \left(\int_{\mathbb{R}} \left| \mathcal{F}\big(1_{A_k}(\cdot) \, \phi(\alpha g(\cdot))\big)(x) \right|^2 dm(x) \right)^{\frac{1}{2}} ; g \in \mathcal{S}, |g| \le |f| \right\}$$

$$= \sup\left\{ \| \mathcal{F}\big(1_{A_k}(\cdot) \, \phi(\alpha g(\cdot))\big) \|_{L^2} ; g \in \mathcal{S}, |g| \le |f| \right\}$$

$$= \sup\left\{ \| 1_{A_k}(\cdot) \, \phi(\alpha g(t)) \|_{L^2} ; g \in \mathcal{S}, |g| \le |f| \right\}$$

$$= \sup\left\{ \left(\int_{A_k} \phi^2\big(\alpha g(t)\big) \, dm(t) \right)^{\frac{1}{2}} ; g \in \mathcal{S}, |g| \le |f| \right\}$$

$$= \left(\int\limits_{A_k} \phi^2 \Big(\alpha f(t) \Big) \, dm(t) \right)^{\frac{1}{2}} = \left(\int\limits_{A_k} \psi \Big(\alpha f(t) \Big) \, dm(t) \right)^{\frac{1}{2}} \rightarrow 0.$$

Hence, $E^\psi \subset E_\rho$. In order to prove the inverse inclusion, let us fix an $\alpha > 0$ and a sequence $A_k \downarrow \emptyset$. For given $f \in E_\rho$, we have then $\rho(\alpha f, A_k) \rightarrow 0$ as $k \rightarrow \infty$. For every simple function $g_k \in \mathcal{S}$ such that $|g_k| \leq |f1_{A_k}|$ we get

$$\left(\int\limits_{\mathbb{R}} \left| \int\limits_{\mathbb{R}} e^{-itx} \, \phi \Big(\alpha g_k(t) \Big) \, dm(t) \right|^2 dm(x) \right)^{\frac{1}{2}} \leq \rho(\alpha f, A_k) \rightarrow 0,$$

which means that $\mathcal{F}\Big(\phi(g_k(\cdot)) \Big) \rightarrow 0$, where $\mathcal{F}$ denotes the Fourier transformation. Since $\mathcal{S} \subset E^\phi$, it follows that the function $\phi\Big(\alpha g_k(\cdot) \Big)$ is a member of the space L^2. The inverse to the Fourier transformation acts continuously from L^2 onto L^2 and therefore the functions $\phi\Big(\alpha g_k(\cdot) \Big)$ converge to zero in L^2, i.e.

$$\int\limits_{\mathbb{R}} \phi^2 \Big(\alpha g_k(t) \Big) \, dm(t) \rightarrow 0.$$

We claim now that $f \in E^\psi$ where $\psi(u) = \Big(\phi(u) \Big)^2$. Indeed, assume to the contrary that

$$\int\limits_{\mathbb{R}} \phi^2 \Big(\alpha f(t) \Big) \, dm(t) = \infty$$

for an $\alpha > 0$. There exist, therefore, $\epsilon > 0$ and a sequence $A_k \downarrow \emptyset$ such that

$$\int_{\mathbb{R}} 1_{A_k}(t)\, \phi^2\Big(\alpha f(t)\Big)\, dm(t) > \epsilon.$$

Thus, we may choose a sequence of simple functions $g_k \in \mathcal{S}$ so that $|g_k| \le |f1_{A_k}|$ and

$$\int_{\mathbb{R}} \phi^2\Big(\alpha g_k(t)\Big)\, dm(t) > \tfrac{\epsilon}{2}.$$

We proved, however, that the integrals

$$\int_{\mathbb{R}} \phi^2\Big(\alpha g_k(t)\Big)\, dm(t) \to 0 \text{ as } k \to \infty.$$

The contradiction completes the proof of the equality $E_\rho = E^\psi$.

6.2.4 EXAMPLE. Let us assume now that $X = \mathbb{R}$, m denotes the Lebesgue measure in $\mathbb{R}$ and $k : \mathbb{R} \times \mathbb{R} \to \mathbb{R}$ is a measurable function such that $k(x,\cdot) \in L^1$ m-a.e. Let ϕ be a function with the same properties as in the previous example. Let us consider the Hammerstein operator of the form

$$T(f)(x) = \int_{\mathbb{R}} k(x,y)\, \phi\Big(f(y)\Big)\, dm(y).$$

Let $H = M = M(X,\mathbb{R})$ be the space of all measurable functions regarded as a topological vector space with the topology of convergence in measure on all subsets of finite measure. This topology may be defined by means of an F-norm

$$\|f\|_m = \int\limits_{\mathbb{R}} \frac{|f(x)|}{1+|f(x)|} \, p(x) \, dm(x),$$

where $p > 0$ m-a.e. is a measurable function such that

$$\int\limits_{\mathbb{R}} p(x) \, dm(x) = 1.$$

Let us define a linear integral operator K by the formula

$$Kf(x) = \int\limits_{\mathbb{R}} k(x,y) \, f(y) \, dm(y).$$

As the domain of this operator we can take the space

$$D_K = \Big\{ f \in M \; ; \int\limits_{\mathbb{R}} |k(x,y)| \, |f(y)| \, dm(y) < \infty \Big\}.$$

The set D_K is usually called the proper domain of the linear operator K. Assume additionally that K is nonsingular, i.e. there exists $g \in D_K$ such that $g > 0$ m-a.e. For every $f \in M$ let us put

$$\|f\|_{\sim} = \|f\|_m + d_K(f)$$

where

$$d_K(f) = \sup\{\|Kg\|_m \; ; \; g \in D_K, \; |g| \le |f| \text{ m-a.e.}\}.$$

It is well known from the theory of linear integral operators (cf. Szeptycki [1], Labuda and Szeptycki [1]) that $\|\cdot\|_{\sim}$ is a complete translation invariant metric on an additive group M. Denoting by $\tilde{D}_K$ the closure in $(M, \|\cdot\|_{\sim})$ of D_K we get the extended domain of the operator K, i.e. K can be extended to a continuous operator $\tilde{K} : \tilde{D}_K \to M$ (c f. Szeptycki [1], Labuda and Szeptycki [1]).

6.2.5 THEOREM. The following two conditions are equivalent:

(a) $f \in E_\rho$, where ρ is a function semimodular given by the operator T;

(b)   The function $\phi\big(\lambda f(\cdot)\big) \in \tilde{D}_K$ for every positive λ.

Before starting the proof of this theorem we have to prove two lemmas.

LEMMA 6.2.6.   For every $f \in M$ and $A \in \Sigma$

$$\rho(f,A) = \sup\left\{\left\|\int_A k(\cdot,y)\,\phi\big(g(y)\big)\,dm(y)\right\|_m\right\},$$

where the supremum is taken over all measurable g such that $|g| \le |f|\,1_A$ m-a.e. and $\phi\big(g(\cdot)\big) \in D_K$.

PROOF.   Observe that if g is a measurable, bounded function then

$$\int_{\mathbb{R}} |k(x,y)|\,|\phi(g(y))|\,dm(y)$$

$$\leq \operatorname*{ess\ sup}_{z \in \mathbb{R}} |\phi(g(z))| \cdot \int_{\mathbb{R}} |k(x,y)| \, dm(y) < \infty$$

for almost all x because $k(x,\cdot) \in L^1(\mathbb{R})$ m-a.e. and therefore $\phi\big(g(\cdot)\big) \in D_K$. Thus,

$$\rho(f,A)$$

$$= \sup\left\{\left\|\int_{\mathbb{R}} k(\cdot,y) \, \phi\big(g(y)\big) \, dm(y)\right\|_m ; g \in \mathcal{S}, |g| \leq |f|1_A\right\}$$

$$\leq \sup\left\{\left\|\int_{A} k(\cdot,y) \, \phi\big(g(y)\big) \, dm(y)\right\|_m ; \phi\big(g(\cdot)\big) \in D_K, |g| \leq |f|1_A\right\}$$

To prove the equality, let us take a function g such that $|g| \leq |f|1_A$ m-a.e. and $\phi\big(g(\cdot)\big) \in D_K$. There exists a sequence (g_k) of $\mathcal{P}$-simple functions such that $|g_k| \uparrow |g|$ m-a.e. Since for every $A \in \Sigma$, we have

$$\int_{A} |k(x,y)| \, |\phi\big(g(y)\big)| \, dm(y) < \infty \quad \text{m-a.e.,}$$

it follows that

$$\lim_{n \to \infty} \int_{A} k(x,y) \, \phi\big(g_n(y)\big) \, dm(y) = \int_{A} k(x,y)\phi\big(g(y)\big) \, dm(y) \quad \text{m-a.e.}$$

Thus

$$\left\| \int_A k(\cdot,y) \, \phi\Big(g_n(y)\Big) \, dm(y) - \int_A k(\cdot,y) \, \phi\Big(g(y)\Big) \, dm(y) \right\|_m \to 0$$

and consequently, to every $\epsilon > 0$ there corresponds $n_0 \in \mathbb{N}$ such that

$$\left\| \int_A k(\cdot,y) \, \phi\Big(g(y)\Big) \, dm(y) \right\|_m \le \epsilon + \left\| \int_A k(\cdot,y) \, \phi\Big(g_{n_0}(y)\Big) \, dm(y) \right\|_m.$$

In view of arbitrariness of $\epsilon > 0$ we get finally

$$\rho(f,A)$$

$$= \sup \left\{ \left\| \int_A k(\cdot,y) \, \phi\Big(g(y)\Big) \, dm(y) \right\|_m \; ; \; \phi\Big(g(\cdot)\Big) \in D_K, \; |g| \le |f| \, 1_A \right\}.$$

6.2.7 LEMMA. For every $f \in M$ and $A \in \Sigma$

$$\rho(f,A)$$

$$= \sup \left\{ \left\| \int_A k(\cdot,y) \, h(y) \, dm(y) \right\|_m \; ; \; h \in D_K, \; |h| \le |\phi(f(\cdot))| 1_A \right\}$$

$$= d_K \Big(\phi(f(\cdot)) \Big).$$

PROOF.   Let   the   functions   h   and   g   be   related   by
$h(y) = \phi\big(g(y)\big)$. Since ϕ is an odd function we have $|h(y)| = \phi\big(|g(y)|\big)$.   On the other hand, ϕ is increasing and therefore the
following statements are equivalent:

$$|h(y)| \le |\phi(f(y))| \, 1_A(y)$$

and

$$|g(y)| \le |f(y)| \, 1_A(y).$$

The rest follows from Lemma 6.2.6.

PROOF of THEOREM 6.2.5.

(a) $\Rightarrow$ (b) Let us take a function $f \in E_\rho$ and arbitrary $\lambda > 0$.

Put

$$Z_n = \{x \in [-n,n] \, ; \, |f(x)| \le n\} \text{ and } A_n = X \setminus Z_n.$$

Thus, $\lambda f 1_{Z_n}$ is measurable and bounded for almost all $x \in \mathbb{R}$ and
therefore

$$\phi\big(\lambda f 1_{Z_n}(\cdot)\big) \in D_K.$$

We have

$$d_K\big(\phi\big(\lambda f(\cdot)\big) - \phi\big(\lambda f 1_{Z_n}(\cdot)\big)\big) = d_K\big(\phi\big(\lambda f(\cdot)1_{A_n}(\cdot)\big)\big) = \rho(\lambda f, A_n),$$

which tends to zero. Since

$$\|\phi(\lambda f(\cdot)) - \phi(\lambda f(\cdot)1_{Z_n}(\cdot))\|_m \to 0,$$

we get

$$\|\phi(\lambda f(\cdot)) - \phi(\lambda f(\cdot)1_{Z_n}(\cdot))\|_{\sim} \to 0,$$

which means that

$$\phi\Big(\lambda f(\cdot)\Big) \in \tilde{D}_K.$$

(b) $\Rightarrow$ (a) Suppose that f is such that $\phi\Big(\lambda f(\cdot)\Big) \in \tilde{D}_K$ for every $\lambda > 0$. Let us fix $\lambda > 0$ and $A_k \downarrow \emptyset$. Since $\phi\Big(\lambda f(\cdot)\Big) \in \tilde{D}_K$, there exists a sequence $g_n \in D_K$ such that $\|g_n - \phi(\lambda f(\cdot))\|_{\sim} \to 0$. In particular, $d_K\Big(g_n - \phi(\lambda f(\cdot))\Big) \to 0$. For a given $\epsilon > 0$ let us fix $n \in \mathbb{N}$ such that

$$d_K\Big(g_n - \phi(\lambda f(\cdot))\Big) < \tfrac{\epsilon}{2}.$$

Observe that

$$d_K(1_{A_k}g_n)$$

$$= \sup\left\{ \left\|\int_A k(\cdot,y)\, g(y)\, dm(y)\right\|_m ; g \in D_K, |g| \le |g_n|1_{A_k} \right\}$$

$$\le \sup\left\{ \left\|\int_A |k(\cdot,y)|\, |g(y)|\, dm(y)\right\|_m ; g \in D_K, |g| \le |g_n|1_{A_k} \right\}$$

$$\le \left\|\int_A |k(\cdot,y)|\, |g(y)|\, dm(y)\right\|_m \to 0,$$

because

$$\int_A |k(\cdot,y)| \, |g(y)| \, dm(y) < \infty \quad \text{m-a.e.}$$

Finally, in view of Lemma 6.2.7

$$\rho(\lambda f, A_k) = d_K\Big(\phi(\lambda f(\cdot))\Big)$$

$$\le d_K\Big(\big(g_n - \phi(\lambda f(\cdot))\big) 1_{A_k}\Big) + d_K\Big(g_n 1_{A_k}\Big)$$

$$\le d_K\Big(g_n 1_{A_k} - \phi(\lambda f(\cdot))\Big) + d_K\Big(g_n 1_{A_k}\Big) < \tfrac{\epsilon}{2} + \tfrac{\epsilon}{2} = \epsilon$$

for k sufficiently large. This means that f belongs to E_ρ which completes the proof.

Theorem 6.2.5 states that a function f belongs to E_ρ, which is a domain of the nonlinear operator

$$T(f)(x) = \int_\mathbb{R} k(x,y) \, \phi\Big(f(y)\Big) \, dm(y)$$

if and only if for every $\lambda > 0$ the function $y \mapsto \phi\Big(\lambda f(y)\Big)$ belongs to the extended domain of the linear operator

$$Kf(x) = \int_\mathbb{R} k(x,y) \, f(y) \, dm(y).$$

Let us observe that if $|\phi|$ satisfies the Δ_2-condition then it suffices to consider the condition $\phi\big(\lambda(\cdot)\big) \in \tilde{D}_K$ for some $\lambda > 0$. Theorem 6.2.5 makes our results comparable with the well developed theory of linear integral operators. Since for particular linear operators K there exist suitable criteria for being a member of the class $\tilde{D}_K$, we can obtain also some convenient characterizations of domains of nonlinear operators. The next result is a good example of this approach.

6.2.8 EXAMPLE. Let T be the nonlinear Fourier transform (see Example 6.2.3)

$$T(f)(x) = \int_{\mathbb{R}} e^{-itx} \, \phi\big(f(t)\big) \, dm(t).$$

By H we denote now the space of all measurable functions M. In view of the results from the preceding section,

$$E_\rho = \Big\{ f \in M \; ; \; \phi\big(\lambda f(\cdot)\big) \in \tilde{D}_{\mathcal{F}} \text{ for every } \lambda > 0 \Big\},$$

where $\mathcal{F}$ is the linear Fourier transformation

$$\mathcal{F}f(x) = \int_{\mathbb{R}} e^{-itx} \, f(t) \, dm(t).$$

It was proved by Szeptycki [1] (see also Labuda, Szeptycki [1]) that $\tilde{D}_{\mathfrak{F}} = l^2(L^1)$. In other words, the domain E_ρ of the operator T consists of all functions f such that

$$\sum_{-\infty}^{\infty} \left(\int_{n}^{n+1} |\phi(\lambda f(y))| \, dm(y) \right)^2 < \infty \quad \text{for every } \lambda > 0.$$

Let us repeat that if $|\phi|$ satisfies the Δ_2-condition then the above becomes equivalent to

$$\sum_{-\infty}^{\infty} \left(\int_{n}^{n+1} |\phi(\lambda f(y))| \, dm(y) \right)^2 < \infty \quad \text{for some } \lambda > 0.$$

6.3 EQUICONTINUITY OF SEQUENCES OF NONLINEAR OPERATORS

Let (T_n) be a sequence of nonlinear operators satisfying (6.1.3), (6.1.4), (6.1.5), (6.1.6) with the same space H; then we can construct the respective function modulars ρ_n and corresponding spaces E_{ρ_n}. Defining

$$\rho(f,A) = \sum_{n=1}^{\infty} 2^{-n} \frac{\rho_n(f,A)}{1 + \rho_n(f,A)},$$

we obtain again a function modular and corresponding space E_ρ such that every $T_n : E_\rho \to H$ is continuous.

Let us note, however, that the sequence (T_n) does not have to be equicontinuous. In order to obtain the equicontinuity of (T_n) one should consider the functional

$$\rho_0(f,A) = \sup_{n \in \mathbb{N}} \rho_n(f,A).$$

It may happen that ρ_0 is not a function modular. One can easily see (cf. Theorem 5.1.3) that ρ_0 is a function modular if the following conditions are satisfied:

(a) $\sup_{n \in \mathbb{N}} \overline{T}_n(\alpha, A_n) \to 0$

for every $\alpha > 0$, $A_n \in \mathcal{P}$, $A_n \downarrow \emptyset$,

(b) $\sup_{n \in \mathbb{N}} \overline{T}_n(\alpha_n, A) \to 0$

for every $A \in \mathcal{P}$, $\alpha_n \geq 0$, $\alpha_n \to 0$.

If (f_k) is a sequence of functions from L_ρ^0 such that $\|f_k\|_\rho \to 0$ as $k \to \infty$, then for every natural n there holds

$$\|T_n(f_k)\|_H \leq \rho_n(f_k) \to 0 \quad \text{as } k \to \infty.$$

The latter fact states simply that every operator T_n is $\|\cdot\|_\rho$-continuous at zero if regarded as an operator acting from L_ρ^0

into H. Furthermore, we can observe that an even stronger continuity holds, namely

$$\|T_n\|_H \to 0 \text{ whenever } \rho(f_n) \to 0 \text{ and } f_n \in L_\rho^0.$$

We may ask the following question: are T_n equicontinuous at zero in that sense? In other words, we are interested in when the following is true:

$$\sup_{n \in \mathbb{N}} \|T_n(f_k)\|_H \to 0 \text{ if } \rho(f_k) \to 0 \text{ and } f_k \in L_\rho^0.$$

Certainly, the above does not hold in general. Let us put, for instance, $H = \mathbb{R}$, $X = [0,1]$, $T_n(f) = \int |f(x)|^n dm(x)$. If we put $f = 2 \cdot 1_X$ we get $f \in L_\rho^0$ while $\sup_{n \in \mathbb{N}} T_n(f) = \infty$.

To the end of the section we will restrict our cosideration to some more special operators T_n. Let $X = \mathbb{R}^+$ and let m denote Lebesgue measure in $\mathbb{R}$. Let us consider the case $H = L^\infty(\mathbb{R}^+)$. Assume that the following estimation holds for every $n \in \mathbb{N}$ and $x \in \mathbb{R}^+$:

$$(6.3.1) \qquad |T_n(f)(x)| \le \int_{\mathbb{R}^+} a(x,y) \, \phi_n\big(|f(y)|\big) \, dm(y),$$

where $a(x,y)$ is a nonnegative measurable function and the functions ϕ_n are continuous, nondecreasing functions acting from $\mathbb{R}^+$ into itself such that $\phi_n(0) = 0$, $\phi_n(u) > 0$ for $u > 0$ and $\phi_n(u) \rightarrow \infty$ as $u \rightarrow \infty$. We seek necessary and sufficient conditions for the equicontinuity of the sequence of dominating operators

$$(6.3.2) \qquad K_n(f)(x) = \int_{\mathbb{R}^+} a(x,y)\ \phi_n\Big(|f(y)|\Big)\ dm(y).$$

Let us note that the Hammerstein operators K_n satisfy conditions (6.1.3) to (6.1.6) and that the function modulars ρ_n induced by K_n are simply given by the formula

$$\rho_n(f,A) = \operatorname*{ess\ sup}_{x \in \mathbb{N}} \int_{\mathbb{R}^+} a(x,y)\ \phi_n\Big(|f(y)|\Big)\ dm(y).$$

The next result is obvious in view of the definition of both modulars ρ and ρ_0.

6.3.3 PROPOSITION. The following are equivalent :

$(6.3.4)$ \qquad The operators K_n are equicontinuous at zero in L_ρ^0;

$(6.3.5)$ \qquad For every sequence of functions (f_k) from L_ρ^0

$$\rho_0(f_k) \rightarrow 0 \ \text{ whenever } \rho(f_k) \rightarrow 0.$$

Modifying slightly the proof of Proposition 5.2.9 one can easily prove the following:

6.3.6 PROPOSITION. Assume that ϕ_k is a regular ϕ_1-function. If for a constant $\lambda_k > 0$ there holds

$$\rho_k(\lambda_k f) = \operatorname*{ess\,sup}_{x \in \mathbb{R}^+} \int_{\mathbb{R}^+} a(x,y)\, \phi_k\left(\lambda_k |f(y)|\right) dm(y) < \infty,$$

then there exists $\delta_k > 0$ such that the function $\delta_k f$ belongs to $L^0_{\rho_k}$.

Using Proposition 6.3.6 and Theorem 5.2.19 it is not difficult to obtain the following characterization of the equicontinuity of K_n.

6.3.7 THEOREM. Let (ϕ_k) be a normal sequence of ϕ_1-functions. If the family of measures given by

$$\mu_x(A) = \int_A a(x,y)\, dm(y) \text{ for } x \in \mathbb{R}^+$$

is equisplittable, order equicontinuous and equibounded then the following conditions are equivalent:

(a) The operators K_n are equicontinuous at zero in L_ρ^0;

(b) The sequence of functions (ϕ_k) is essentially finite.

Let us observe that if (ϕ_k) is an essentially finite sequence of regular ϕ-functions there exists then $i_0 \in \mathbb{N}$ such that ϕ_i is equivalent to ϕ_{i_0} for every $i \geq i_0$, which implies the equivalence of modulars ρ_i and ρ_{i_0}. The latter means simply that for every sequence $(f_k) \subset L_\rho^0$ and every $i \geq i_0$ the following statement is true:

$$\|K_i(f_k)\|_H \to 0 \text{ if and only if } \|K_{i_0}(f_k)\|_H \to 0.$$

Therefore, the meaning of Theorem 6.3.7 is as follows: the construction of the modular ρ certainly guarantees the equicontinuity of any finite number of operators K_n but one should not expect any interesting result for an infinite sequence of operators.

6.4 SOME FIXED POINT THEOREMS

Let us discuss now some problems connected with fixed point theory. As we know, the modular function space L_ρ is a metric space; in particular it is an F-space. We can use, therefore, any fixed point theorems which are expressed in terms of metric or F-spaces. Since the F-norm $\|\cdot\|_\rho$ is given in implicit form, however, it may be very difficult or even impossible to verify the assumptions of such theorems. **That is why we are interested**, as usual in this book, in obtaining results which can be formulated in modular terms. Let us start with a result which could be called the Banach Contraction Principle for modular function spaces.

6.4.1 THEOREM. Let ρ be a function modular satisfying the Δ_2-condition and let B be a closed subset of L_ρ such that

$$\sup\{\rho(f - g) \; ; \; f, g \in B\} = R < \infty.$$

If $T : B \rightarrow B$ is a strict ρ-contractive nonlinear operator, i.e. there exists $c < 1$ such that for all functions f and g from B there holds:

$$(6.4.2) \qquad \rho\Big(T(f) - T(g)\Big) \leq c \, \rho(f - g),$$

then there exists a unique function $\widetilde{f} \in B$ such that $T(\widetilde{f}) = \widetilde{f}$.

PROOF. Let us observe that for every f, g $\in$ B there holds:

$$\rho\Big(T^n(f) - T^n(g)\Big) \leq c^n \, \rho(f - g),$$

where $T^n(f)$ denotes the n-th iterate of $T(f)$. In particular, for every function $f_0 \in B$ and every natural numbers k and n we have

$$\rho\Big(T^{k+n}(f_0) - T^n(f_0)\Big) \leq c^n \ \rho\Big(T^k(f_0) - f_0\Big) \leq c^n R \rightarrow 0.$$

Since ρ satisfies the Δ_2-condition, then the above implies that the sequence $\Big(T^n(f_0)\Big)$ is a Cauchy sequence in L_ρ. The metric space L_ρ is complete and there exists therefore a function $\widetilde{f} \in L_\rho$ such that

$$\rho\Big(T^n(f_0) - \widetilde{f}\Big) \rightarrow 0.$$

The function $\widetilde{f}$ belongs to B because B is a closed subset of L_ρ. Let us observe that $\widetilde{f}$ is a fixed point for T. Indeed,

$$\rho\Big(T(\widetilde{f}) - T^n(f_0)\Big) \leq c \ \rho\Big(\widetilde{f} - T^{n-1}(f_0)\Big) \rightarrow 0.$$

But

$$\rho\Big(\widetilde{f} - T^n(f_0)\Big) \rightarrow 0$$

as well and consequently

$$T(\widetilde{f}) = \widetilde{f} \quad \rho\text{-a.e.}$$

Uniqueness follows from

$$\rho(\widetilde{f}_1 - \widetilde{f}_2) = \rho\Big(T(\widetilde{f}_1) - T(\widetilde{f}_2)\Big) \leq c \ \rho(\widetilde{f}_1 - \widetilde{f}_2),$$

and since $c < 1$, $\rho(\widetilde{f}_1 - \widetilde{f}_2) < \infty$, this is possible only when $\widetilde{f}_1 = \widetilde{f}_2 \quad \rho$-a.e.

In order to study existence of fixed points in L_ρ for ρ without Δ_2 we have to introduce a new property of function modulars.

6.4.3 DEFINITION. We say that the function modular ρ satisfies the (∗)-condition if and only if

$$\rho(\widetilde{f},H) \leq \limsup_{n \to \infty} \rho(f_n,H)$$

for every $H \in \mathcal{P}$ and $f_n \Rightarrow \widetilde{f}$ on H.

Let us note that the (∗)-condition holds for a very large class of function modulars. For instance, if ρ has the Fatou property then ρ satisfies (∗). Indeed, if $H \in \mathcal{P}$, and f_n converges uniformly to f on H, then by the Fatou property we get

$$\rho(f,H) = \rho(f1_H) \leq \liminf_{n \to \infty} \rho(f_n 1_H) \leq \limsup_{n \to \infty} \rho(f_n,H).$$

6.4.4 LEMMA. Let ρ satisfy the (∗)-condition. If f_n, $f \in E_\rho$ and there exists a subsequence (g_n) of (f_n) with $g_n \to f$ ρ-a.e. then for all $g \in E_\rho$

$$\rho(f - g) \leq \limsup_{n \to \infty} \rho(f_n - g).$$

PROOF. By the Egoroff Theorem there exists a sequence $H_k \uparrow X$, $H_k \in \mathcal{P}$ and g_n converges uniformly to f on every H_k. For fixed $k \in \mathbb{N}$ we have, via the $(*)$-condition,

$$\rho(f - g, H_k) \leq \limsup_{n \to \infty} \rho(g_n - g, H_k)$$

$$\leq \limsup_{n \to \infty} \rho(g_n - g) \leq \limsup_{n \to \infty} \rho(f_n - g).$$

Since $f - g \in E_\rho$ we get

$$\rho(f - g, X \setminus H_k) \to 0 \text{ as } k \to \infty.$$

and finally

$$\rho(f - g, H_k) \to \rho(f - g) \text{ as } k \to \infty.$$

This implies

$$\rho(f - g) \leq \limsup_{n \to \infty} \rho(f_n - g)$$

as claimed.

6.4.5. THEOREM. Assume that ρ satisfies the $(*)$-condition. Let B be a closed subset of E_ρ and B be sequentially compact in the sense of convergence ρ-a.e. If $\sup\{\rho(f - g); f, g \in E_\rho\} = R < \infty$ and $T : B \to B$ is a strict ρ-contraction then there exists a unique fixed point $\tilde{f} \in B$.

PROOF. Let us take an arbitrary function $f_o \in B$ and put $f_n = T^n(f_o)$, and define

$$\phi(y) = \limsup_{n \to \infty} \rho(f_n - g).$$

Since T is a strict contraction, we obtain $\phi\big(T(g)\big) \leq c\phi(g)$ for all $g \in B$, where $c < 1$ is the Lipschitzian constant for T. Thus,

$$\inf_{g \in B} \phi(g) = 0.$$

By the ρ-a.e. compactness of B we can choose a subsequence (g_n) of (f_n) such that $g_n \to \tilde{f}$ with $f \in B$. By Lemma 6.4.4 $\rho(\tilde{f} - g) \leq \phi(g)$ for any $g \in B$. We have

$$\rho\big(\tilde{f} - T^n(g)\big) \leq \phi\big(T^n(g)\big) \leq c^n \phi(g)$$

and

$$\rho\big(T(\tilde{f}) - T^n(g)\big) \leq \rho\big(\tilde{f} - T^{n-1}(g)\big) \leq c\phi\big(T^{n-1}(g)\big) \leq c^n\phi(g).$$

Finally,

$$\rho\Big(\tfrac{1}{2}\big(\tilde{f} - T(\tilde{f})\big)\Big) \leq \rho\big(\tilde{f} - T^n(g)\big) + \rho\big(T^n(g) - T(\tilde{f})\big)$$

$$\leq 2c^n \phi(g) \leq 2c^n R \to 0.$$

Hence,

$$T(\tilde{f}) = \tilde{f} \quad \rho\text{-a.e.}$$

Before we pass to the fixed point theorem for ρ-nonexpansive mappings, i.e. satisfying (6.4.2) with the constant $c = 1$, we have to define the following function

$$w_r(t,A) = \sup\left\{\frac{\rho(tf,A)}{\rho(f,A)} \; ; f \in L_\rho, \, 0 < \rho(f) \leq r\right\},$$

where $t \in [0,1]$, $r \in [0,\infty]$, $A \in \Sigma$.

Note that $w_r(t,A) \leq 1$, $w_r(0,A) = 0$ and $w_r(1,A) = 1$. It is also easy to see that if ρ is s-convex $(0 < s \leq 1)$ then $w_r(t,A) < 1$ for all $t < 1$.

6.4.6 DEFINITION. A set $B \subset L_\rho$ is called a star-shaped set if there exists a center of B, that is, such a function $u \in B$ that $u + \lambda(f - u)$ belongs to B for every $\lambda \in (0,1)$ and every $f \in B$.

6.4.7 THEOREM. Let B be a star-shaped subset of E_ρ, sequentially compact in the sense of convergence ρ-a.e. Assume that for every $r \in \mathbb{R}^+$, $A \in \Sigma$ and every $t \in [0,1)$, $w_r(t,A) < 1$. Assume also that B is ρ-bounded, i.e.

$$\sup\{\rho(f - g) \; ; f, g \in B\} = R < \infty.$$

Let us assume that the function modular ρ satisfies the (∗)-condition and that for every ρ-a.e. convergent sequence of functions (f_n) from B, $\rho(f_n, \cdot)$ are order equicontinuous. If T: B → B is ρ-nonexpansive, i.e.

$$\rho\left(T(f) - T(g)\right) \le \rho(f - g);$$

then there exists $\widetilde{f} \in$ B such that

$$T(\widetilde{f}) = \widetilde{f} \ \rho\text{-a.e.}$$

PROOF. For every $\lambda \in (0,1)$ let us consider the operator $T_\lambda : B \to B$ defined by

$$T_\lambda(f) = u + \lambda\left(T(f) - u\right),$$

where $u \in B$ is a center of B. Observe that

$$\rho\left(T_\lambda(f) - T_\lambda(g)\right) = \rho\left(\lambda\left(T(f) - T(g)\right)\right)$$

$$\le w_R(\lambda, X) \, \rho\left(T(f) - T(g)\right) \le w_R(\lambda, X) \, \rho(f - g)$$

and, because $w_R(\lambda, X) < 1$, T_λ is strictly ρ-contractive. By Theorem 6.4.5 there exists a fixed point $f_\lambda \in$ B, i.e. $T_\lambda(f_\lambda) = f_\lambda$. Let us take a sequence of positive numbers $\lambda_n \uparrow 1$. Denoting $T_n = T_{\lambda_n}$, $f_n = f_{\lambda_n}$ and using the compactness argument we can assume (passing to a suitable subsequence if necessary) that there exists a measurable function $\widetilde{f} \in B$ such that $f_n \to \widetilde{f}$ ρ-a.e. By the Egoroff Theorem there exists a sequence (H_k) of sets from $\mathcal{P}$ such that $H_k \uparrow X$ and f_n converges uniformly to the function $\widetilde{f}$ on every H_k, hence $\rho\left(\alpha(f_n - \widetilde{f}), H_k\right) \to 0$ for every $\alpha > 0$ and $k \in \mathbb{N}$. Observe that

(i) $\qquad \rho\left(\tfrac{1}{2}(T(\widetilde{f}) - T_n(\widetilde{f})), H_k\right) \leq \rho\left(T(\widetilde{f}) - T_n(\widetilde{f})\right)$

$$= \rho\left((1 - \lambda_n)\,(T(\widetilde{f}) - u)\right) \to 0,$$

because the function $T(\widetilde{f}) - u$ belongs to L_ρ and $1 - \lambda_n \to 0$. Moreover, we have

(ii) $\qquad \rho\left(T_n(f_n) - f_n, H_k\right) \leq \rho\left(T_n(f_n) - f_n\right) = 0,$

because f_n is a fixed point for T_n. Observe then that

(iii) $\qquad \rho\left(T_n(\widetilde{f}) - T_n(f_n), H_k\right) \leq \rho\left(T_n(\widetilde{f}) - T_n(f_n)\right)$

$$\leq \rho\left(\lambda_n\left(T(\widetilde{f}) - T(f_n)\right)\right) \leq \rho\left(T(\widetilde{f}) - T(f_n)\right)$$

$$\leq \rho(\widetilde{f} - f_n) \leq \rho(\widetilde{f} - f_n, H_m) + \rho(\widetilde{f} - f_n, X \setminus H_m)$$

holds for all natural numbers n, k, m. Hence, for all n, k, m we have

(iv) $$\rho\Big(T_n(\tilde{f}) - T_n(f_n), H_k\Big)$$

$$\leq \rho(\tilde{f} - f_n, H_m) + \sup_{n \in \mathbb{N}} \rho(\tilde{f} - f_n, X \setminus H_m).$$

Since $X \setminus H_m \downarrow \emptyset$, for arbitrary $\epsilon > 0$ we can find m_0 such that

$$\sup_{n \in \mathbb{N}} \rho(\tilde{f} - f_n, X \setminus H_{m_0}) < \frac{\epsilon}{2}.$$

On the other hand, (f_n) converves uniformly to $\tilde{f}$ on H_{m_0} and consequently

$$\rho(\tilde{f} - f_n, H_{m_0}) < \frac{\epsilon}{2} \text{ for n sufficiently large.}$$

Thus we have proved that for every $k \in \mathbb{N}$

(v) $$\lim_{n \to \infty} \rho\Big(T_n(\tilde{f}) - T_n(f_n), H_k\Big) = 0.$$

By (i), (ii) and (v), for arbitrary $k \in \mathbb{N}$

(vi) $$\rho\Big(\frac{1}{4}\big(T(\tilde{f}) - f_n, H_k\big)\Big)$$

$$\leq \rho\Big(\frac{1}{2}\big(T(\tilde{f}) - T_n(\tilde{f})\big), H_k\Big) + \rho\Big(T_n(\tilde{f}) - T_n(f_n), H_k\Big)$$

$$+ \rho\Big(T_n(f_n) - f_n, H_k\Big) \to 0 \text{ as } n \to \infty.$$

Finally,

$$\rho\Big(\tfrac{1}{8}\big(T(\tilde{f}) - \tilde{f}\big), H_k\Big)$$

$$\leq \rho\Big(\tfrac{1}{4}\big(T(\tilde{f}) - f_n\big), H_k\Big) + \rho\Big(\tfrac{1}{4}(\tilde{f} - f_n), H_k\Big) \to 0,$$

which implies that $T(\tilde{f}) = \tilde{f}$ ρ-a.e. in H_k. Since $H_k \uparrow X$ we conclude that $T(\tilde{f}) = \tilde{f}$ ρ-a.e. in X. The theorem is completely proved.

Let us observe that Theorem 6.4.7 covers in particular the case of the space $L^1[0,1]$. Alspach [1] produced an example of a weakly compact convex subset B of $L^1[0,1]$ and of a nonexpansive operator $T : B \to B$ which was fixed point free. In this sense our result which uses the concept of compactness in the sense of ρ-a.e. convergence can be helpful in the situations where the more classical, weak compactness arguments do not work. This is possible because we did not use any geometrical properties of modular function spaces. If a modular function space is a uniformly convex Banach space then the classical results of Browder [1], Göhde [1] and Kirk [1] may be certainly applied.

Bibliographical remarks

In the case of linear integral transformations the question of extending the domain was thoroughly elaborated (the reader is referred to the papers of Aronszajn and Szeptycki [1], Szeptycki [1], [2], Labuda and Szeptycki [1], [2]). Similar results for nonlinear operators are not familiar to the author. The possibility of describing the domain of nonlinear operators in terms of extended domains of some associated linear operators (see Example 6.2.4, Theorem 6.2.5 and Example 6.2.8) was suggested by Szeptycki [3]. The question of extending the domain of the Nemytskii operator from a closed ball in L^p to the entire space L^p was raised by Krasnosel'skii in his book [1]; these results were then generalized by Kozlowski [1], [2], [3] to the case of a larger class of nonlinear operators acting between some Banach function spaces. To extend the domain of T from $\mathcal{E}$ we used, however, a different method because we constructed the domain dependent upon the operator itself. In a different context a similar concept was employed by Jedryka and Musielak [1] (see also Musielak [1], Section 21) in order to obtain some existence theorems for integral equations. The general method of construction described above follows some suggestions contained in Kozlowski [5], Example 8.5. These suggestions were based on the theory of nonlinear operator measures (cf. Batt [1],

Friedman and Tong [1], *Kozlowski and Szczypinski* [2], [3], *Szczypiński* [1]). *Theorem 6.1.9 and some of the examples were given in the paper* [6] *of Kozlowski. The construction of the extended domains has been a subject for further generakization (Labuda* [1]). *In the case of Lipschitzian operators one can consider the problem of extension (preserving the Lipschitz constant) of such operators from a set D to its closed convex hull. In the Banach space setting this problem was elaborated by Wells and Williams in their book* [1], *and for the accretive sets in Banach spaces, by Reich* [1]. *For a general review of fixed point theory the reader is referred to the monographs of Dugundji and Granas* [1], *Goebel and Reich* [1], *Kirk* [2,3]. *Fixed point theorems using the modular conditions rather than metric ones and employing sequential compactness in the sense of convergence in measure were given by Lami Dozo and Turpin* [1] *in 1987 for the case of Musielak-Orlicz spaces. In the case of modular function spaces it is also possible to consider the notion of ρ-normal structure and obtain in this way some nonconstructive fixed point theorems for ρ-nonexpansive mappings (cf. Khamsi, Kozlowski, Reich* [1]). *If L_ρ is an Orlicz space, one can prove some fixed point theorems for ρ-nonexpansive mappings without any compactness arguments and all additional assumptions on the space are expressed entirely in terms of simple properties of an Orlicz function ϕ (see Khamsi, Kozlowski, Shutao* [1]).

7. Application to Approximation Theory

7.1 PRELIMINARIES

Let X be a compact subset of C^n. Throughout this chapter, we will consider the case of complex valued functions, i.e. $S = C$. Let μ be a finite Borel measure on X. Let $\Sigma = \mathcal{P}$ be a σ-algebra of Borel subsets of X. We assume, therefore, that L_ρ is a subspace of all the μ-measurable functions M(X), ρ being a function modular. In this chapter we consider the problem of analytic extension of a function $f \in L_\rho$ to a holomorphic function $\tilde{f}$ defined on an open neighborhood of X. Let us start with the following definition.

7.1.1 DEFINITION. Let X be a Borel subset of C^n and let μ be a Borel measure on X. We say that the pair (X, μ) satisfies the L^*-condition (the L^*-condition at a point $a \in \bar{X}$) if and only if for every family $\mathcal{F}$ of polynomials such that

$$\mu\{t \in X \sup_{p \in \mathcal{F}} |p(t)| = +\infty\} = 0$$

and for every $b > 1$ there exists $M > 0$ and U an open

neighborhood of $\overline{X}$ (open neighborhood of a) such that

$$\sup_{t \in U} |p(t)| \leq M \, b^{\deg p} \quad \text{for every } p \in \mathcal{F}.$$

It is clear that if X is a compact subset of $\mathbb{C}^n$ then the pair (X,μ) satisfies the L^*-condition if and only if the pair (X,μ) satisfies the L^*-condition at each point $a \in X$. We will give some examples of pairs (X,μ) satisfying the L^*-condition.

7.1.2 EXAMPLE. If X is a rectifiable Jordan arc in $\mathbb{C}$ and μ is a measure of length over X then (X,μ) satisfies L^* at every point $a \in X$.

By Fubini's Theorem, from Example 7.1.2 we get the next example.

7.1.3 EXAMPLE. Let X be a subset of the space $\mathbb{R}^n$ (R^n is treated as a subset of $\mathbb{C}^n$ such that $\mathbb{C}^n = \mathbb{R}^n + i\mathbb{R}^n$). Let m_n denote the Lebesgue n-dimensional measure. (X,m_n) satisfies L^* at $a \in X$ if there exists a nonsingular affine mapping $l : I_n \to \mathbb{R}^n$ such that $a \in l(I_n) \subset X \cup \{a\}$, where I_n is the n-th Cartesian power of $I = [0,1]$. In particular, for every bounded convex set $X \subset \mathbb{R}^n$ such that $int(X) \neq \emptyset$ or else for every bounded Lipschitz domain (of class Lip 1) the pair (X,m_n) satisfies L^* at every point $a \in \overline{X}$. We want to add that the condition L^* is invariant under nondegenerate holomorphic mappings from C^n to C^m ($m \leq n$). We have also the following geometrical criterion for L^*:

7.1.4 THEOREM. (Pleśniak [3]) Given $a \in \bar{X}$; suppose that there exists an analytic mapping $h : [0, 1] \rightarrow \bar{X}$ such that $h(0) = a$. The pair (X,μ) satisfies L^* at $a \in \bar{X}$ if the pair (X,μ) satisfies L^* at $h(t)$ for each $t \in (0, 1]$.

In the sequel we will also need the following concept.

7.1.5 DEFINITION. Let X be a compact subset of C^n. Put

$$\hat{X} = \{t \in C^n : |p(t)| \leq \|p\|_X \text{ for every polynomial p}\};$$

$\|p\|_X$ denotes the supremum norm on the set X. We say that X is polynomially convex if $\hat{X} = X$.

In the case when X is a compact subset of C, it follows from the Runge Theorem that X is polynomially convex if and only if $C \cup \{\infty\} \setminus X$ is connected.

Let us introduce the notation:

(1)   By H(X) we denote the class of all complex valued functions on X that admit holomorphic extension to some open neighborhoods of X.

(2)   Put $H_\rho(X) = H(X) \cap L_\rho$.

(3) P_k will stand for the class of all complex polynomials in n variables of degree less than or equal to k.

(4) $\text{dist}_{\|\cdot\|_X}(f, P_k) = \inf\{\|f - p\|_X \; ; \; p \in P_k\}$.

We need the following version of the Bernstein-Walsh Theorem (cf. Siciak [1], [2]).

7.1.6 THEOREM.   Assume the set $X \subset C^n$ is compact and polynomially convex. If $f \in H(X)$ then

$$\limsup_{k \to \infty} \, [\text{dist}_{\|\cdot\|_X}(f, P_k)]^{\frac{1}{k}} < 1.$$

Our next, technical result will be frequently used.

7.1.7 LEMMA. Let ρ be a function modular and let $f_k \in L_\rho$. If there exists a constant $c > 1$ such that

$$\rho(c^k f_k) \to 0$$

$$\left(\text{resp.} \sum_{k=1}^{\infty} \rho(c^k f_k) < +\infty, \quad \limsup_{k \to \infty} \, [\rho(c^k f_k)]^{\frac{1}{k}} < 1 \right),$$

then for every $b \in (1, c)$

$$\|b^k f_k\|_\rho \to 0$$

$$\left(\text{resp.} \sum_{k=1}^{\infty} \|b^k f_k\|_\rho < +\infty, \quad \limsup_{k \to \infty} [\|b^k f_k\|_\rho]^{\frac{1}{k}} < 1 \right).$$

PROOF. Fix $b \in (1, c)$ and put

$$d_k = \max\left\{ \left(\frac{b}{c}\right)^k, \rho(c^k f_k) \right\}.$$

Then

$$\rho\left(\frac{b^k f_k}{d_k} \right) \le \rho(c^k f_k) \le d_k$$

which gives $\|b^k f_k\|_\rho \le d_k$. The rest of the proof is elementary. In the case of s-convex ρ we can obtain similar result for $\|\cdot\|_\rho^s$.

7.2 HOLOMORPHIC EXTENSION OF FUNCTIONS

By the ρ-distance from a measurable function f to the set P_k we understand the quantity

$$\text{dist}_\rho(f, P_k) = \inf\{\rho(f - p) \, ; \, p \in P_k\}.$$

In general, the ρ-distance is not a distance in the metric sense and can be infinite.

First, we are going to establish a sufficient condition for being a member of the class $H_\rho(X)$.

7.2.1 THEOREM. Let X be a compact, polynomially convex subset of $\mathbf{C}^n$. If $f \in H_\rho(X)$ then there exists a constant $c > 1$ such that

$$(7.2.1.a) \qquad \lim_{k \to \infty} \mathrm{dist}_\rho(c^k f, P_k) = 0.$$

PROOF. In view of 7.1.6, for every function f in $H_\rho(X)$ there holds

$$\limsup_{k \to \infty} [\mathrm{dist}_{\|\cdot\|_X}(f, P_k)]^{\frac{1}{k}} < 1.$$

This means that there exists $k_0 \in \mathbb{N}$ and a constant $b < 1$ such that $\|f - p_k\|_X \leq b^k$ for some $p_k \in P_k$ ($k \geq k_0$). Let $c > 1$ be such that $cb < 1$. By the definition of the ρ-distance and by P2, P4 we get

$$\mathrm{dist}_\rho(c^k f, P_k) \leq \rho\left(c^k(f - p_k)\right)$$

$$\leq \rho\left(\|c^k(f - p_k)\|_X, X\right) \leq \rho\left(c^k\|f - p_k\|_X, X\right)$$

$$\leq \rho(c^k b^k, X) \leq \rho\left((cb)^k, X\right) \to 0 \text{ as } k \to \infty.$$

Let us recall that $\rho(\alpha,X) = \sup\{\rho(r1_X); |r| \leq \alpha\}$ and $X \in \mathcal{P}$. By the use of the Taylor expansion and the properties of Siciak's extremal function of the set X (cf. Siciak [2], Plesniak [2]) one can prove that a function is entire in $\mathbb{C}^n$ if and only if for every compact set $X \subset \mathbb{C}^n$ there holds

$$\lim_{k \to \infty} [\text{dist}_{\|\cdot\|_X} (f,P_k)]^{\frac{1}{k}} = 0.$$

Using the above it is not hard to modify the proof of Theorem 7.2.1 in order to obtain our next result.

7.2.2 REMARK. Let X be a compact subset of $\mathbb{C}^n$ and let f be an entire function. Then for every $c > 1$,

$$\lim_{k \to \infty} \text{dist}_\rho(c^k f, P_k) = 0.$$

Under an additional hypothesis we can reformulate Theorem 7.2.1 and obtain a version very similar to the Bernstein-Walsh Theorem.

7.2.3 REMARK. Assume that X is compact and polynomially convex. Assume furthermore that for every $a \in (0, 1)$

$$\limsup_{k \to \infty} [\rho(a^k, X)]^{\frac{1}{k}} < 1.$$

If $f \in H_\rho(X)$ then

(7.2.3.a) $$\limsup_{k \to \infty} [\text{dist}_{\|\cdot\|_\rho}(f, P_k)]^{\frac{1}{k}} < 1.$$

PROOF. Using the estimation from Theorem 7.2.1 we have

$$\rho\Big(c^k(f - P_k)\Big) \leq \rho(a^k, X), \quad \text{where } a = cb < 1.$$

By (7.2.3.a) and Lemma 7.1.7,

$$\limsup_{k \to \infty} [\|f - P_k\|_\rho]^{\frac{1}{k}} < 1,$$

as claimed.

One may ask whether the condition (7.2.1.a) is also sufficient for $f \in H_\rho(X)$. It will turn out (Theorem 7.2.17) that sometimes it is the case. In the general situation, however, we have to assume more. Namely, we have the following :

7.2.4 THEOREM. Assume that the pair (X,μ) satisfies L^* and μ is absolutely continuous with respect to ρ (i.e. μ equals zero on ρ-null sets). Let $f \in L_\rho$. If there exist $d > 1$ and an increasing sequence $k_i \in \mathbb{N}$ such that

(7.2.4.a) $$k_i \to +\infty, \quad \limsup_{i \to \infty} \frac{k_{i+1}}{k_i} < \infty \quad \text{and}$$

$$\sum_{i=1}^{\infty} \mathrm{dist}_\rho \left(d^{k_i} f, P_{k_i} \right) < \infty,$$

then $f \in H_\rho(X)$.

PROOF. By Lemma 7.1.7 we can choose $c \in (1, d)$ such that

$$\sum_{i=1}^{\infty} \mathrm{dist}_{\|\cdot\|_\rho} \left(c^{k_i} f, P_{k_i} \right) < \infty.$$

Put

$$d_j = \mathrm{dist}_{\|\cdot\|_\rho} \left(c^{k_j} f, P_{k_j} \right) + 2^{-j} \quad \text{for } j = 1, 2, \ldots .$$

For every $j \in \mathbb{N}$ we can then pick up a polynomial $p_j \in P_{k_j}$ such that

$$\| c^{k_j}(f - p_j) \|_\rho < d_j.$$

Observe that

$$\| c^{k_j}(p_{j+1} - p_j) \|_\rho \leq \| c^{k_j}(f - p_{j+1}) \|_\rho + \| c^{k_j}(f - p_j) \|_\rho$$

$$\leq \| c^{k_{j+1}}(f - p_{j+1}) \|_\rho + \| c^{k_j}(f - p_j) \|_\rho \leq d_{j+1} + d_j.$$

Since $d_j \to 0$, there exists j_0 such that $d_{j+1} + d_j < 1$ for $j \geq j_0$. From the above we conclude then that

$$\| c^{k_j}(p_{j+1} - p_j) \|_\rho < 1 \quad \text{for } j \geq j_0,$$

which yields the following inequality

$$\rho\left(c^{k_j}(p_{j+1} - p_j)\right) \leq \|c^{k_j}(p_{j+1} - p_j)\|_\rho \ .$$

Finally,

$$\rho\left(c^{k_j}(p_{j+1} - p_j)\right) \leq d_{j+1} + d_j \ \text{ for } j \geq j_0.$$

Let us define the set

$$D = \left\{ t \in X \ ; \ \sup_{j \in N} \ c^{k_j} \ |p_{j+1}(t) - p_j(t)| = +\infty \right\}.$$

We want to prove that $\mu(D) = 0$. For every $j, n \in N$ let

$$X_{j,n} = \left\{ t \in X \ ; \ c^{k_j}|p_{j+1}(t) - p_j(t)| > n \right\} \ \text{ and } \ X_n = \bigcup_{j=n}^{\infty} X_{j,n}.$$

Clearly (X_n) decreases and $\displaystyle\bigcap_{n=1}^{\infty} X_n = D$. Let us fix $\alpha > 0$; for $n \geq \alpha$ and $j \geq j_0$ we have

$$\rho(\alpha, X_{j,n}) \leq \rho(n, X_{j,n}) \leq \rho\left(c^{k_j}(p_{j+1} - p_j), X_{j,n}\right)$$

$$\leq \rho\left(c^{k_j}(p_{j+1} - p_j)\right) \leq d_j + d_{j+1}.$$

Thus,

$$\rho(\alpha, X_n) \leq \sum_{j=n}^{\infty} \rho(\alpha, X_{j,n})$$

$$\leq \sum_{j=n}^{\infty} d_j + \sum_{j=n}^{\infty} d_{j+1} \to 0$$

as $n \to \infty$, because $\displaystyle\sum_{j=1}^{\infty} d_j < \infty$. Since $D \subset X_n$ for every $n \in N$, it follows that

$$\rho(\alpha, D) \leq \rho(\alpha, X_n) \to 0$$

and consequently

$$\rho(\alpha, D) = 0 \quad \text{for all } \alpha > 0,$$

which means that D is ρ-null and, therefore, $\mu(D) = 0$, since $\mu \ll \rho$. By the L^*-condition, for every $b > 1$ there exist $M > 0$ and an open neighborhood U of X such that

$$\sup_{t \in U} \left| c^{k_j}\big(p_{j+1}(t) - p_j(t)\big) \right| \leq b^{k_{j+1}} \cdot M \qquad \text{for } j = 1, 2, \dots .$$

Since $c > 1$ and $\displaystyle\limsup_{j \to \infty} \frac{k_{j+1}}{k_j} < \infty$ we may choose

$$k > \limsup_{j \to \infty} \frac{k_{j+1}}{k_j} \quad \text{and } b > 1 \text{ such that } b < c^{1/k}.$$

Observe that $\dfrac{k_{j+1}}{k_j} < 1$ for large j and then $c^{1/k} < c^{k_j/k_{j+1}}$.
Hence,

$$\sup_{t \in U} |p_{j+1}(t) - p_j(t)| \leq M \cdot \frac{b^{k_{j+1}}}{c^{k_j}}$$

$$= M \cdot \frac{b^{k_{j+1}}}{c^{(k_j/k_{j+1}) \cdot k_{j+1}}} \leq M \cdot \left(\frac{b}{c^{1/k}} \right)^{k_{j+1}}$$

for j sufficiently large. Since $b < c^{1/k}$ the series

$$p_1 + \sum_{j=1}^{\infty} (p_{j+1} - p_j)$$

is uniformly convergent in U to a holomorphic function $\tilde{f}$. This implies that $\|p_j 1_X - \tilde{f} 1_X\|_\rho \to 0$. Since $\|p_j - f\|_\rho \to 0$, it follows that $f = \tilde{f} 1_X$ ρ-a.e., which gives $f = f 1_X$ μ-a.e. by the absolute continuity of μ with respect to ρ. This completes the proof.

To the end of this section we will study various conditions that can sometimes replace (7.2.4.a).

7.2.5 LEMMA. The following conditions are equivalent:

(7.2.6)   There exists $K > 1$ such that $\rho(2f) \leq K\rho(f)$ for every $f \in$

(7.2.6)   To every $b > 1$ there exist $c > 1$ and $k_0 \in \mathbb{N}$ with
$\rho(c^k f) < b^k \rho(f)$ for all $f \in L_\rho$ and every $k \geq k_0$.

PROOF. It is easy to show by induction that if ρ satisfies (7.2.6) then $\rho(2^n f) \leq K^n \rho(f)$ for every $f \in L_\rho$ and $n \in \mathbb{N}$. Let us fix $b > 1$. If $b \geq K$ then we can put $c = 2$ and $k_0 = 1$. Aassume then that $b \in (1,K)$ and choose $k_0 \in \mathbb{N}$ such that $b^{k_0} \geq K$. Let us put $c = 2^{1/2k_0}$ and fix $k \geq k_0$. We can find m and p in $\mathbb{N}$ such that $m \geq p$, $p < k_0$ and $k = mk_0 + p$.

Compute:

$$b^k \rho(f) \geq b^{mk_o} \rho(f) \geq K^m \rho(f) \geq \rho(2^m f) =$$

$$= \rho(c^{2mk_o} f) \geq \rho(c^{mk_o+p} f) = \rho(c^k f).$$

To prove the converse, fix an arbitrary $b > 1$. By (7.2.7) we get an index $k_o \in N$ and a constant $c > 1$ such that

$$\rho(c^k f) \leq b^k \rho(f) \quad \text{for} \quad k \geq k_o, f \in L_\rho.$$

Choose $k_1 \geq k_o$ such that $c^{k_1} \geq 2$. Thus,

$$\rho(2f) \leq \rho(c^{k_1} f) \leq b^{k_1} \rho(f),$$

which completes the proof.

Lemma 7.2.5 will be used to prove the next result, in which we replace the convergence of the series in (7.2.4.a) by a more convenient condition .

7.2.8 THEOREM. Let μ and X be as in the hypothesis of Theorem 7.2.4. Assume additionally that ρ satisfies (7.2.6). If there exists a strictly increasing sequence (k_j) of natural numbers such that

(7.2.9) $\lim_{j\to\infty} \sup \dfrac{k_{j+1}}{k_j} < \infty$ and $\lim_{j\to\infty} \sup [\text{dist}_\rho(f, P_{k_j})]^{\frac{1}{k_j}} < 1,$

then $f \in H_\rho(X)$.

PROOF. In view of Theorem 7.2.4 it suffices to find a constant $c > 1$ such that

$$\sum_{j=1}^{\infty} \text{dist}_\rho\left(c^{k_j}f, P_{k_j}\right) < \infty.$$

By (7.2.9) we can pick a number $d \in (0, 1)$ such that

$$\text{dist}_\rho\left(f, P_{k_j}\right) \leq d^{k_j} \text{ for j sufficiently large.}$$

Let us choose $b > 1$ that $bd < 1$. Then

$$b^{k_j}\text{dist}_\rho(f, P_{k_j}) < (db)^{k_j}$$

and by Lemma 7.2.5 we may find $c > 1$ such that

$$\text{dist}_\rho\left(c^{k_j}f, P_{k_j}\right) \leq b^{k_j}\text{dist}_\rho\left(f, P_{k_j}\right) < (bd)^{k_j}.$$

Thus, the series

$$\sum_{j=1}^{\infty} \text{dist}_\rho\left(c^{k_j}f, P_{k_j}\right)$$

converges, which completes the proof.

7.2.10 REMARK. Let us note that the F-norm $\|\cdot\|_\rho$ induced by an arbitrary function modular ρ may be regarded itself as a function modular. By the triangle inequality $\|\cdot\|_\rho$ satisfies (7.2.6). We may replace, therefore, (7.2.9) by

$$(7.2.11) \qquad \limsup_{j \to \infty} \frac{k_{j+1}}{k_j} < \infty, \text{ and}$$

$$\limsup_{j \to \infty} \left[\text{dist}_{\|\cdot\|_\rho} \left(f, P_{k_j} \right) \right]^{\frac{1}{k_j}} < 1.$$

The philosophy of this book is that it is easier to verify conditions expressed in terms of the function modular ρ than conditions involving the corresponding F-norm $\|\cdot\|_\rho$ because of the implicit definition of the latter. It is worth noting, however, that in some cases both methods are equivalent.

7.2.12 PROPOSITION. If ρ satisfies (7.2.6) then (7.2.9) is equivalent to (7.2.11).

PROOF. Since $\rho(f) \leq \|f\|_\rho < 1$ holds, it follows that (7.2.11) always implies (7.2.9). To prove the converse, let us choose $d \in (0, 1)$ such that

$$\text{dist}_\rho(f, P_k) < d^k \text{ for } k \geq k_o.$$

There exists a sequence of $w_k \in P_k$ such that $\rho(f - w_k) < d^k$.

Let $b > 1$ be so chosen that $bd < 1$. By Lemma 7.2.7 there exists $c > 1$ and $k_o \in \mathbb{N}$ such that for $k \geq k_o$

$$\rho\left(c^k(f - w_k)\right) \leq b^k \rho(f - w_k) \leq (bd)^k \to 0 \text{ as } k \to \infty.$$

The rest of the proof will be divided into two parts.

(a)   Let ρ be s-convex ($s \in (0,1]$). Since $\rho\left(c^k(f - w_k)\right) \leq 1$ for large k, then

$$\|f - w_k\|_\rho^s \leq \left(\tfrac{1}{c^k} \right)^s = \left(\tfrac{1}{c^s} \right)^k.$$

Thus,

$$[\text{dist}_{\|\cdot\|_\rho^s} (f, P_k)]^{\frac{1}{k}} \leq \left(\|f - w_k\|_\rho^s \right)^{\frac{1}{k}} \leq \tfrac{1}{c^s} < 1.$$

Hence ,

$$\limsup_{k \to \infty} \; [\text{dist}_{\|\cdot\|_\rho}(f, P_k)]^{\frac{1}{k}} \leq \tfrac{1}{c^s} < 1.$$

(b)   Suppose ρ is not s-convex. Put $q = \max\left(bd, \tfrac{1}{c}\right)$; then we have

$$\rho\left(\frac{f - w_k}{q_k} \right) \leq q^k$$

and consequently

$$\|f - w_k\|_\rho \leq q^k.$$

Thus ,

$$\limsup_{k \to \infty} \ [\text{dist}_{\|\cdot\|_\rho} \ (f, \ P_k)]^{\frac{1}{k}} \leq q < 1.$$

As an example of a function modular satisfying (7.2.6) one can take the Orlicz modular induced by a ϕ-function with the Δ_2-property.

7.2.13 PROPOSITION. Assume the following:

(7.2.14)   To every b > 1 there corresponds d > 0 such that $\|bf\|_\rho \geq (1 + d)\|f\|_\rho$ for all $f \in L_\rho$.

If there exists c > 1 with

$$\text{dist}_\rho(c^k f, \ P_k) \to 0$$

then

$$\limsup_{k \to \infty} \ [\text{dist}_{\|\cdot\|_\rho}(f, \ P_k)]^{\frac{1}{k}} < 1.$$

PROOF. By Lemma 7.1.7 we can choose $b \in (1,c)$ such that

$$\text{dist}_{\|\cdot\|_\rho} \ (b^k f, \ P_k) \to 0$$

and consequently

$$\|b^k(f - p_k)\|_\rho \to 0 \ \text{ for some } p_k \in P_k.$$

By (7.2.14) for k sufficiently large

$$\|f - p_k\|_\rho (1 + d)^k \leq \|b^k(f - p_k)\|_\rho < 1$$

which gives the desired result.

Let us note that every s-homogenous norm satisfies (7.2.14) with $1 + d = c^s$.

To obtain a sufficient condition for (7.2.14) in terms of function modulars we define the following function

$$w(t) = w_\infty(t) = \sup \left\{ \frac{\rho(tf)}{\rho(f)} \; ; f \in L_\rho \backslash \{0\} \right\}.$$

Recall that this function was already used in Section 6.4.

7.2.15 PROPOSITION. If to every $c > 1$ there corresponds $d > 0$ such that

$$(7.2.16) \qquad\qquad w\left(\frac{1 + d}{c}\right) \cdot (1 + d) \leq 1,$$

then (7.2.14) holds.

PROOF. Fix $c > 1$ and choose $d > 0$ such that (7.2.16) holds.

Put

$$L = \frac{1}{1+d} \cdot \left\{ \alpha > 0 \; ; \; \rho\left(\frac{cf}{\alpha}\right) \le \alpha \right\}$$

and

$$P = \left\{ \beta > 0 \; ; \; \rho\left(\frac{f}{\beta}\right) \le \beta \right\}.$$

We claim that $L \subset P$. Indeed, let $\gamma \in L$; then

$$\rho\left(\frac{f}{\gamma}\right) = \rho\left(\frac{\frac{1+d}{c}cf}{\gamma(1+d)} \right) \le w\left(\frac{1+d}{c} \right) \rho\left(\frac{cf}{\gamma(1+d)} \right)$$

$$\le w\left(\frac{1+d}{c} \right) (1+d)\gamma \le \gamma,$$

i.e. $\gamma \in P$. We have

$$\frac{1}{1+d} \, \|cf\|_\rho = \inf L \ge \inf P = \|f\|_\rho.$$

Thus, the condition (7.2.14) holds.

7.2.17 THEOREM. Let μ and X be as in the hypothesis of Theorem 7.2.4. Let $f \in L_\rho$ and let us assume additionally that ρ satisfies (7.2.16) (in particular ρ can be s-convex or satisfy (7.2.14)). If $\lim\limits_{k \to \infty} \mathrm{dist}_\rho (c^k f, P_k) = 0$ then $f \in H_\rho(X)$.

PROOF. In view of Proposition 7.2.15 the condition (7.2.16) implies (7.2.14). By Proposition 7.2.13 we have then

$$\limsup_{k \to \infty} \ [\text{dist}_{\|\cdot\|_\rho}(f, P_k)]^{\frac{1}{k}} < 1,$$

which by Remark 7.2.10 and Theorem 7.2.8 implies that $f \in H_\rho(X)$.

7.2.18 EXAMPLE. Let $X \subset \mathbb{C}$ be a circle of radius $r \in (0,1)$ and center at zero. Consider a function f defined by the formula

$$f(x) = \sum_{k=0}^{\infty} a_k \, f_k(x),$$

where $a_k \in \mathbb{C}$, f_k are measurable functions defined in $\mathbb{C}$ and the convergence of the series is understood in the sense of the $\|\cdot\|_\rho$−norm. Assume $I_k \subset X$ are measurable, $f_k(x) = x^k$ for every $x \in I_k$ and $|a_k| \leq M$ for all $k \in \mathbb{N}$. We would like to find out when f can be extended to a holomorphic function in a neighborhood of X. Let μ be the Lebesgue measure in $\mathbb{C}$ and ρ be a function modular satisfying (7.2.16) such that μ is absolutely continuous with respect to ρ. It is well known that (X,μ) satisfies the L^*-condition in this case. In view of Theorem 7.2.17 it suffices to find $d > 1$ such that

$$\text{dist}_\rho(d^k f, P_k) \to 0 \ \text{ as } k \to \infty.$$

Denoting

$$w_k = \sum_{i=0}^{k} a_i x^i \text{ and } D_k = X \setminus I_k,$$

we get

$$\text{dist}_\rho\left(d^k f, P_k\right) \le \rho\left(d^k(f - w_k)\right)$$

$$\le \rho\left(d^k(f - w_k), I_k\right) + \rho\left(d^k(f - w_k), D_k\right)$$

$$\le \rho\left(d^k \cdot \sum_{i=k+1}^{\infty} |a_i| r^i, X\right) + \rho\left(d^k(f - w_k), D_k\right)$$

$$\le \rho\left(\frac{Mr}{1-r} \cdot (dr)^k, X\right) + \rho\left(d^k(f - w_k), D_k\right).$$

The left-hand side term tends to 0 for $dr < 1$. The question of analytic extension of f has been, therefore, reduced to the question of whether $\rho(d^k(f - w_k), D_k)$ tends to zero. The latter problem may be solved in various ways depending on the form of the modular. For instance, let

$$\rho(f) = \int_X \varphi(|f|) \, d\mu$$

be an Orlicz modular with φ satisfying the Δ_2-condition. Observe that there exists a function $g \in L_\rho$ such that for $k \in \mathbb{N}$ sufficiently large,

$$|f(x) - w_k(x)| \le |g(x)| \quad \text{for all } x \in I_k,$$

then

$$\rho\left(d^k(f - w_k), D_k\right) = \int_{D_k} \varphi(d^k|f - w_k|) \, d\mu$$

$$\le \tilde{M}^k \int_{D_k} \varphi(|f - w_k|) \, d\mu \le \tilde{M}^k \int_{D_k} \varphi(|g|) \, d\mu,$$

where $\tilde{M}$ is the Δ_2-constant. We conclude then, that in order to prove $f \in H_\rho(X)$ it suffices to check whether $\int_{D_k} \varphi(|g|)\, d\mu$ tends to zero faster than $\tilde{M}^k$ tends to infinity.

7.3 QUASIANALYTIC FUNCTIONS

This section will be devoted to studying some relations between modular function spaces and quasianalytic functions in the sense of Bernstein. Let us start with the following definition.

7.3.1 DEFINITION. Given a Borel subset X of the space $\mathbb{C}^n$, a function $f \in L_\rho$ is said to be $\|\cdot\|_\rho$-quasianalytic on X if there exists an increasing sequence (k_j) of positive integers and a sequence of polynomials (p_j) with $\deg(p_j) \leq k_j$ $(j \in \mathbb{N})$ such that

$$(7.3.2) \qquad \limsup_{j \to \infty} \left[\|f - p_j\|_\rho \right]^{\frac{1}{k_j}} < 1.$$

It is interesting to study when a quasianalytic function belongs to $H_\rho(X)$. By Remark 7.2.10 this holds if the pair (X,μ) satisfies the condition L^* and

$$\limsup_{j \to \infty} \frac{k_{j+1}}{k_j} < \infty.$$

In Theorem 7.3.8 it will be shown that this assumption on the sequence (k_j) cannot be removed. First, we have to introduce some new notions.

7.3.3 DEFINITION. Let ρ be a function modular and $Z \subset L_\rho$. Define

$$R(Z) = \inf\{r(f) \; ; f \in Z \setminus \{0\}\},$$

where

$$r(f) = \sup\{\|tf\|_\rho \; ; t \geq 0\}.$$

Similarly we may put

$$R_\rho(Z) = \inf\{r_\rho(f) \; ; f \in Z \setminus \{0\}\},$$

where

$$r_\rho(f) = \sup\{\rho(tf) \; ; t \geq 0\}.$$

7.3.4 PROPOSITION. Let Z be an arbitrary subset of $L_\rho \setminus \{0\}$; then $R(Z) = 0$ if and only if $R_\rho(Z) = 0$.

PROOF. ($\Rightarrow$) Assume $R(Z) = 0$; fix $\epsilon \in (0, 1)$ and choose a function $f \in Z \setminus \{0\}$ such that $r(f) < \epsilon$. Then, we have

$$r_\rho(f) = \sup\{\rho(tf);\ t \geq 0\} \leq \sup\{\|tf\|_\rho;\ t \geq 0\} = r(f) < \epsilon$$

and by arbitrariness of ϵ we get $R_\rho(Z) = 0$.

$(\Rightarrow)$ If $R_\rho(Z) = 0$ then to every $\epsilon > 0$ there corresponds a function $f \in Z \setminus \{0\}$ such that

$$\rho\left(\frac{t}{\epsilon}\, f \right) < \epsilon \text{ for every } t \geq 0.$$

Hence, $\|tf\|_\rho < \epsilon$ for arbitrary $t \geq 0$ and consequently $R(Z) = 0$.

The following version of Bernstein's "lethargy" theorem will be used in order to obtain a characterization of those modular function spaces in which there exist quasianalytic functions that cannot be extended to holomorhic functions.

7.3.5 THEOREM. (Lewicki [1]) Let $(X, \|\cdot\|)$ be an F-space and let

$$V_1 \subsetneq V_2 \subsetneq \cdots \subsetneq X$$

be a nested sequence of distinct, finite-dimensional vector subspaces of X. Assume that

$$R\left(\bigcup_{n=1}^{\infty} V_n \right) > 0.$$

Then, for every $0 \le d_n \downarrow 0$ there exists $x \in X$ such that

$$\text{dist}_{\|\cdot\|} (x, V_n) = d_n$$

for n sufficiently large.

7.3.6 THEOREM. Let X is a polynomially convex, compact subset of C^n, ρ is a function modular such that

$$R\left(\bigcup_{k=1}^{\infty} P_k\right) > 0.$$

If (k_j) is a sequence of natural numbers such that

(a)
$$\limsup_{j \to \infty} \frac{k_{j+1}}{k_j} = \infty,$$

(b)
$$\limsup_{k \to \infty} \left[\rho\left(a^k, X\right)\right]^{\frac{1}{k}} < 1 \quad \text{for every } a \in (0, 1),$$

then there exists a $\|\cdot\|_\rho$-quasianalytic function $f \in L_\rho \setminus H_\rho(X)$.

PROOF. We note that by the assumption on the sequence (k_j) there is a subsequence (k_{j_m}) of (k_j) such that

$$\lim_{m \to \infty} \frac{k_{j_{m+1}}}{k_{j_m}} = \infty.$$

For simplicity we let (k_j) be the subsequence. Let us fix arbitrary $b \in (0,1)$. By Theorem 7.3.5 there exists a function $f \in L_\rho$ such that for $k \geq k_0$

$$\text{dist}_{\|\cdot\|_\rho} (f, P_k) = d_k,$$

where $d_k = b^{k_j}$ for $k \in [\, k_j \,, k_{j+1}), j = 1, 2, \ldots$. We note that

$$\lim_{j \to \infty} \left[\text{dist}_{\|\cdot\|_\rho}(f, P_{k_j}) \right]^{\frac{1}{k_j}} = \lim_{j \to \infty} \left(b^{k_j} \right)^{\frac{1}{k_j}} = b < 1,$$

while

$$\limsup_{k \to \infty} \left[\text{dist}_{\|\cdot\|_\rho}(f, P_k) \right]^{\frac{1}{k}}$$

$$\geq \limsup_{j \to \infty} \left[\text{dist}_{\|\cdot\|_\rho}\left(f, P_{k_{j+1}-1}\right) \right]^{\frac{1}{-1+k_{j+1}}}$$

$$\geq \lim_{j \to \infty} b^{\frac{k_j}{-1+k_{j+1}}} = 1.$$

Since ρ satisfies (b), by Remark 7.2.3 the function f does not belong to $H_\rho(X)$.

Let us observe that the hypothesis of Theorem 7.3.6 is always satisfied if ρ is s-convex and $\rho(1,X) < \infty$. Since for s-convex function modulars we have always $R(Z) > 0$ for all nontrivial sets $Z \subset L_\rho$ then the class of modular function spaces for which $B_\rho \setminus H_\rho \neq \emptyset$ is quite large, where by B_ρ we denote the clas of all quasianalytic functions in L_ρ. It contains for instance all s-convex Musielak-Orlicz spaces $L^\phi(X,\Sigma,\mu)$ where μ is a finite Borel measure on a compact, polynomially convex subset of $\mathbb{C}^n$.

Below, we consider two general conditions that imply in particular $R(L_\rho) = 0$ and $R(Z) > 0$ for nontrivial Z.

7.3.7. DEFINITION.

(a) We say that ρ satisfies (C_1) if and only if
$$\lim_{n \to \infty} \sup \{\rho(\alpha, A_n); \alpha \geq 0\} = 0 \text{ as}$$
$A_n \downarrow \emptyset, A_n \in \Sigma.$

(b) ρ satisfies (C_2) if there exists a constant $d > 0$ such that
$\sup\{\rho(\alpha, A); \alpha \geq 0\} \geq d$
for every $A \in \Sigma$ which is not ρ-null.

7.3.8 PROPOSITION.

(a) If ρ satisfies (C_1) and there exists a set $X_1 \subset X$ such that ρ is atomless in X_1 then $R_\rho(L_\rho) = 0$.

(b) If ρ satisfies (C_2) then $R_\rho(L_\rho) > 0$.

PROOF. (a) Since ρ is atomless in X_1 we can choose a sequence (A_n) of Σ-measurable subsets of X_1 such that $A_n \downarrow \emptyset$

and A_n is not ρ-null for $n \in \mathbb{N}$. Put

$$V_n = \{f \in S \; ; \; f1_{X \setminus A_n} = 0\}.$$

Observe that V_n are nontrivial subspaces of L_ρ. For every
$f \in V_n$ there holds

$$r_\rho(f) = \sup\{\rho(tf); \; t \geq 0\}$$

$$= \sup\{\rho(tf, A_n) \; ; \; t \geq 0\} \leq \sup\{\rho(\alpha, A_n); \; \alpha \geq 0\}.$$

Hence,

$$R_\rho(L_\rho) = \inf\{r_\rho(f); \; f \in L_\rho \setminus \{0\}\}$$

$$\leq \inf\{r_\rho(f); \; f \in \bigcup_{n=1}^{\infty} V_n \setminus \{0\}\} \leq \sup\{\rho(\alpha, A_n); \; \alpha \geq 0\}$$

for $n = 1, 2, \dots$, and consequently $R_\rho(L_\rho) = 0$.

(b)   Let $Z \neq \{0\}$; take any function $f \in Z \setminus \{0\}$ and denote

$$A_n = \{x \in X \; ; \; |f(x)| \geq \tfrac{1}{n}\}.$$

The sequence (A_n) is nondecreasing and A_n are not ρ-null for
$n \geq n_0$. Then for $n \geq n_0$ we have

$$r_\rho(f) \geq \rho(n^2 f) \geq \rho(n^2 f, A_n) \geq \rho(n, A_n) \geq \rho(n, A_{n_0}).$$

Hence,

$$r_\rho(f) \geq \sup\{\rho(\alpha, A_{n_o}) ; \alpha \geq 0\} \geq d$$

and consequently

$$R_\rho(Z) \geq d > 0.$$

As an example of a function modular with property (C_2) one can take the Orlicz modular or Musielak-Orlicz modular, provided in the last case that for every set A of positive measure,

$$\sup_{\alpha>0} \int_A \phi(\alpha,x) \, d\mu(x) = +\infty.$$

The space L_ρ with $R(L_\rho)=0$ seems to be rather awkward since it contains arbitrarily short lines (in the terminology of Rolewicz [1]). It is not surprising then that (C_1) looks very strange too. It holds, however, for the modular of the form

$$\rho(f,A) = \int_A \phi\left(|f(x)|\right) d\mu(x),$$

where μ is finite and atomless and the function ϕ satisfies $\lim\sup_{u \to \infty} \phi(u) < \infty$. One can easily see that L_ρ contains in this case all measurable functions.

It can be shown (cf. Pleśniak [2], Proposition 1.4) that, if the set of all polynomials is dense in L_ρ, then the modular function space L_ρ coincides with $B_\rho(X) + {}_{\circ}B_\rho(X)$. This is an interesting property because in $B_\rho(X)$ we have a strong identity principle (see Theorem 7.3.9 below). This principle can be proved for arbitrary connected $X \subset \mathbb{R}^n$ with nonempty interior. For the sake of simplicity we will prove it only for compact intervals. The reader is referred to the paper [2] of Pleśniak for further details.

7.3.9 THEOREM. Assume that X is a compact interval included in $\mathbb{R}^n$ and μ is a Borel measure on X. Assume furthermore that for each compact interval $I = [a_1, b_1] \times \cdots \times [a_n, b_n]$ contained in X the pair (I,μ) satisfies L^*. Let $f \in B_\rho(X)$. If f vanishes on a subset F of X such that (F,μ) satisfies L^* then $f = 0$ μ-a.e. in X.

PROOF. Since $f \in B_\rho(X)$, there exist a strictly increasing sequence of integers (k_j), polynomials (p_j) with $\deg(p_j) \leq k_j$ and a constant $a \in (0, 1)$ such that

$$\|f - p_j\|_\rho \leq a^{k_j} \text{ for j sufficiently large.}$$

We note that $\|p_j 1_F\|_\rho = \|(f - p_j)1_F\|_\rho \leq \|f - p_j\|_\rho$. Thus, for large j, $\|p_j 1_F\|_\rho \leq a^{k_j}$. Choose $b > 1$ such that $ab < 1$. Since each F-norm satisfies condition (7.2.6), by Lemma 7.2.5, there exists $c \in (1,b)$ such that

$$(7.3.10) \qquad \left\| c^{k_j} p_j 1_F \right\|_\rho \le b^{k_j} \left\| p_j 1_F \right\|_\rho < \left(ab \right)^{k_j}.$$

Put

$$D = \left\{ t \in F \; ; \sup_{j \in N} \left| c^{k_j} p_j(t) \right| = +\infty \right\}.$$

We claim that $\mu(D) = 0$. To do this, define for j and n in N,

$$F_{n,j} = \left\{ t \in F; \; \left| c^{k_j} p_j(t) \right| > n \right\} \text{ and } F_n = \bigcup_{j=n}^{\infty} F_{n,j}.$$

Using (7.3.10), by similar reasoning as in Theorem 7.2.4, we can show that for every $\alpha > 0$,

$$\lim_{n \to \infty} \rho(\alpha, F_n) = 0$$

which gives that $\rho(\alpha, D) = 0$ and consequently $\mu(D) = 0$. Now choose $d \in (1,c)$. By condition L^* there exists a closed subinterval I_O of X and a constant $M > 0$ such that

$$\sup_{t \in I_O} |p_j(t)| \le M \left(\tfrac{d}{c} \right)^{k_j} \text{ for j sufficiently large.}$$

Hence, the sequence (p_j) tends uniformly to 0 in I_O which implies that $f = 0$ μ-a.e. on I_O. Let J_O be a maximal element of the family $\mathcal{J}$ of all compact subintervals I of X such that $I_O \subset I$ and $f = 0$ μ-a.e. on I. We claim that $J_O = X$. Indeed, since $f = 0$ μ-a.e. on J_O we have

$$|p_j 1_{J_O}| \le a^{k_j} \text{ for j sufficiently large}$$

and since (J_o, μ) satisfies L^* we can choose again a compact interval J such that $J_o \subset$ int (J) and p_j tends uniformly to 0 in J as $j \to \infty$. Thus $f = 0$ μ-a.e. on $J \subset X$, hence $J \cap X = J_o$. This is, however, possible only if $J_o = X$ as claimed.

Let us note that if a function $f \in L_\rho$ satisfies (7.3.2) then there also holds

$$(7.3.11) \qquad \limsup_{j \to \infty} \rho(f - p_j)^{\frac{1}{k_j}} < 1,$$

where (k_j) is an increasing sequence of positive integers and p_j are polynomials of degree not greater than k_j. This observation suggests that we may introduce another weaker version of quasianalyticity.

7.3.12 DEFINITION. Let X be a Borel subset of $\mathbb{C}^n$. A function $f \in L_\rho$ is called ρ-quasianalytic on X if and only if there exists an increasing sequence of natural numbers (k_j) and a sequence of polynomials (p_j) with deg $(p_j) \leq k_j$ such that (7.3.11) holds.

Clearly every $\|\cdot\|_\rho$-quasianalytic function is ρ-analytic as well. Theorem 7.3.6 states that it can happen that there are

$\|\cdot\|_\rho$-quasianalytic functions which cannot be extended to holomorphic functions. We will present now an analogous result for the case of ρ-quasianalytic functions; we can get rid of some assumptions of Theorem 7.3.6 and replace them by more natural conditions. In order to do this let us recall the definition of the Δ'_2-modulars (Def. 3.3.1). The function modular ρ was said to satisfy the Δ'_2-condition if to every $d > 0$ there exists $c(d) > 0$ such that $\rho(f + g) \le c(d)$ whenever $\max\big(\rho(f),\ \rho(g)\big) \le d$. We will use the following version of the Bernstein Theorem.

7.3.13 THEOREM. Let ρ be a continuous, Δ'_2-function modular and let

$$V_1 \subsetneq V_2 \subsetneq \cdots \subsetneq X$$

be a nested sequence of distinct, finite-dimensional vector subspaces of X. Assume that

$$R_\rho\Big(\bigcup_{n=1}^{\infty} V_n \Big) > 0.$$

Then, for every $d_0 > 0$ such that the set $\{f \in V_n;\ \rho(f) < d_0\}$ is compact for $n \in \mathsf{N}$ and for every $0 \le d_n \downarrow 0$, $c(d_1) < d_0$, there exists $f \in L_\rho$ such that $\mathrm{dist}_\rho (f, V_n) = d_n$ for all $n \in \mathsf{N}$.

7.3.14 THEOREM. Let X be a polynomially convex, compact subset of $\mathbb{C}^n$ and let $(k_j) \subset \mathbb{N}$ be such that

$$\limsup_{j \to \infty} \frac{k_{j+1}}{k_j} = \infty.$$

Assume that a function modular ρ satisfies the Δ'_2-condition and

$$R_\rho \left(\bigcup_{k=1}^\infty P_k \right) > 0 \, .$$

There exists then a ρ-quasianalytic function f which does not belong to $H_\rho(X)$.

PROOF. Similarly as in the proof of 7.3.6 we may fix a $\in (0,1)$ and assume that

$$\mathrm{dist}_\rho(f, P_k) = d_k \text{ where } d_k = a^{k_j} \text{ for } k \in [k_j, k_{j+1}), j \in \mathbb{N}.$$

Since

$$\lim_{j \to \infty} \left[\mathrm{dist}_\rho(f, P_{k_j}) \right]^{\frac{1}{k_j}} = \lim_{j \to \infty} \left(a^{k_j} \right)^{\frac{1}{k_j}} = a < 1,$$

it follows that f is ρ-quasianalytic. On the other hand, for every $c > 1$ there holds

$$\limsup_{k \to \infty} \left[\mathrm{dist}_\rho(c^k f, P_k) \right]^{\frac{1}{k}} \geq \limsup_{k \to \infty} \left[\mathrm{dist}_\rho(f, P_k) \right]^{\frac{1}{k}}$$

$$\geq \lim_{j \to \infty} \sup \left[\text{dist}_\rho\left(f, P_{k_{j+1}-1}\right) \right]^{\frac{1}{-1+k_{j+1}}}$$

$$\geq \lim_{j \to \infty} a^{\frac{k_j}{-1+k_{j+1}}} = 1.$$

In view of Theorem 7.2.1, f does not belong to $H_\rho(X)$.

By Theorem 3.3.15, in the case when ρ is of finite type, and such that $L_\rho^j \subset L_\rho^0$, the condition Δ_2' may be replaced by Δ_2.

Bibliographical remarks

The idea of expressing some extension properties by means of polynomial approximation has its origin in results of S.N. Bernstein from the 1930's (cf. Bernstein [1]). The notion of L^*-condition was studied extensively by Pleśniak (see his papers [2] and [3]); the last paper contains also a more complete list of examples. The polynomial lemma of Leja was first published in paper [1] of Leja in 1933. Theorem 7.1.4 was taken from

Plesniak [3] *while the version of the Bernstein-Walsh Theorem* (*Theorem 7.1.6*) *was given by Siciak* [1]. *The first results related to the question of extension of measurable functions were given by Plesniak* [2], [3] *for the case of classical Orlicz spaces* L^ϕ *for convex and* Δ_2*-functions* ϕ. *The case of modular function spaces was considered by Kozlowski and Lewicki in their papers* [1], [2]. *A current review of the theory of quasianalytic functions in the sense of S.N. Bernstein can be found in Plesniak* [1]. *The versions of the Bernstein Theorem used in this chapter were taken from Lewicki* [1].

Bibliography

J. Albrycht and J. Musielak

[1]   Countably modulared spaces, Studia Math. 31
       (1968), 331-337.

D.E. Alspach

[1]   A fixed point free nonexpansive map, Proc. Amer.
       Math. Soc. 82(1981), 423 - 424.

N. Aronszajn and P. Szeptycki

[1]   On general integral transformations, Math. Ann. 163
       (1966), 127-154.

J. Batt

[1]   Nonlinear integral operators on C(S,E), Studia Math.
       48.2 (1973), 145-177.

S.N.Bernstein

[1]   Collected works, A.N. SSSR, Moscow 1952 - 1954
       (in Russian).

Z. Birnbaum and W. Orlicz

[1]   Über die Verallgemeinerung des Begriffes der
       zueinander konjugierten Potenzen,
       StudiaMath. 3 (1931), 1-67.

F.E. Browder

[1]   Nonexpansive nonlinear operators in a Banach space,
       Proc. Nat. Acad. Sci. U.S.A. 54(1965), 1041 - 1044.

I. Dobrakov

[1]   On integration in Banach spaces I, Czech. Math. J.
       20 (1970), 511-536.
[2]   On integration in Banach spaces II, Czech. Math. J.
       20 (1970), 680-695.

L. Drewnowski

[1]   Topological rings of sets, continuous set functions,
       integration I, Bull. Acad. Polon. Sci.
       Ser. Sci. Math. Astronom. Phys. 20 (1972), 269-276.
[2]   Topological rings of sets, continuous set functions,
       integration II, Bull. Acad. Polon. Sci.
       Ser. Sci. Math. Astronom. Phys. 20 (1972), 277-286.

L. Drewnowski and A. Kamińska

[1]   Orlicz spaces of vector functions generated by a
       family of measures, Comment. Math. 22
       (1981), 175-186.

J. Dugundji and A. Granas

[1]   Fixed point theory, Vol.1. PWN Warszawa 1982.

N. Dunford and J. Schwartz

[1]   Linear Operators I, Interscience, New York 1958.

N. Friedman and A.E. Tong

[1]   On additive operators, Canad. J. Math. 23 (1971),
       468-480.

K. Goebel and S. Reich

[1]   Uniform convexity, hyperbolic geometry, and
       nonexpansive mappings. Dekker 1984,
       Monographs and textbooks in pure and applied
       mathematics 83.

D. Göhde

[1]   Zum Prinzip der kontraktiven Abbildung, Math.
       Nachr. 30(1965), 251-263.

P.R. Halmos

[1]   Measure Theory, D. Van Nostrand, New York 1956.

H. Hudzik

[1]   Convexity in Musielak-Orlicz spaces, Hokkaido Math.
        J. 14 (1985), 271-284.

H. Hudzik, J. Musielak and R. Urbański

[1]   Some extensions of the Riesz-Thorin theorem to
        generalized Orlicz spaces $L_M^\mu(T)$,
        Comment. Math. 22 (1980), 43-61.

[2]   The Riesz-Thorin theorem in generalized Orlicz
        spaces of nonsymmetric type, Publ. Elektr.
        Fak. Univ. u Beogradu, 706 (1980), 145-158.

[3]   Interpolation of compact sublinear operators in
        generalized Orlicz spaces of nonsymmetric type,
        Topology, Vol.I, Proceedings of the 4thColloqium,
        Budapest 1978, Coll. Math. Soc. Ja. Bolyai 23
        (1980), 625-638.

T.M. Jedryka and J. Musielak

[1]   On a modular equation, Functiones et Approximatio
        3(1976), 101-111.

A. Kamińska

[1]   Thesis, Poznań 1978.

[2]   On some compactness criteria for Orlicz subspace
      $E_\Phi(\Omega)$, Comment. Math. 22 (1982), 245-255.

[3]   Strict convexity of sequence Orlicz-Musielak spaces
      with Orlicz norm, J. Funct. Anal. 50(1983), 285-305.

[4]   On uniform convexity of Orlicz spaces, Indagationes
      Mathematicae 44.1 (1982), 27-36.

M.A. Khamsi, W.M. Kozlowski, S. Reich

[1]   Fixed point theory in modular function spaces
      (to be published).

M.A. Khamsi, W.M. Kozlowski, Ch. Shutao

[1]   Some geometrical properties of Orlicz spaces
      (to be published).

W.A. Kirk

[1]   A fixed point theorem for mappings which do not

increase distances, Amer. Math. Monthly
72(1965), 1004 - 1006.

[2]   Fixed point theory for nonexpansive mappings,
Lecture Notes in Mathematics, Vol. 886,
Berlin, Heidelberg, New York, 1981, 484 - 505.

[3]   Fixed point theory for nonexpansive mappings II,
Contemporary Math., 18(1983), 121 - 140.

A. Kozek

[1]   Orlicz spaces of functions with values in Banach
spaces, Comment. Math. 19 (1976), 259-288.

[2]   Convex integral functionals on Orlicz spaces,
Comment. Math. 21 (1979), 109-135.

W.M. Kozlowski

[1]   Nonlinear operators in Banach function spaces,
Comment. Math. 22 (1980), 85-103.

[2]   A note on the continuity of nonlinear operators, Coll.
Math. Soc. Jan. Bolyai 35 (1980), 745-750.

[3]   Boundedness of G-dominated nonlinear operators,

Approximation and Function Spaces,Ed. Z. Ciesielski,
PWN Warszawa, North Holland Amsterdam, New
York, Oxford, 1981.

[4]   Notes on modular function spaces I, Comment. Math.
        28 (1988), 91-104.

[5]   Notes on modular function spaces II, Comment.
        Math. 28 (1988), 105-120.

[6]   On domains of some nonlinear operators (to appear
        in Journal of Mathematical Analysis and
        Applications).

W.M. Kozlowski and G. Lewicki

[1]   On polynomial approximation in modular function
        spaces, Proceedings of the International Conference
        "Function Spaces" held in Poznań, August 1986, Ed.
        J. Musielak (to be published in Lecture Notes in
        Mathematics, Springer Verlag).

[2]   Analyticity and polynomial approximation in
        modular function spaces, (to appear in Journal of
        Approximation Theory).

W.M. Kozlowski and T. Szczypiński

[1] A note on the Hammerstein operator in Köthe spaces, Acta Math. Univ.Jagiell. 22 (1980),147-150.

[2] Some remarks on the nonlinear operator measures and integration, Coll.Math. Soc. Jan. Bolyai 35 (1980), 751-756.

[3] Convergence theorems for integrals with respect to nonlinear operator measures, Constructive Function Theory 1981, Sofia 1983, 389-392.

P. Kranz and W. Wnuk

[1] On the representation of Orlicz lattices, Indagationes Math. 43.4 (1981), 375-383.

M.A. Krasnosel'skii

[1] Topological methods in the theory of nonlinear equations, Pergamonn Press, Oxford, London, New York, Paris 1974.

M.A. Krasnosel'skii and Ya. B. Rutickii

[1]   Convex functions and Orlicz spaces, P. Noordhoff,
        Ltd., Groningen 1961.

M. Krbec

[1]   Modular interpolation spaces, Zeitschrift f. Anal. u. i.
        Anwen. 1 (1982), 25-40.

I. Labuda

[1]   Personal communication, 1988.

I. Labuda and P. Szeptycki

[1]   Extended domains of some integral operators with
        rapidly oscillating kernels, Indagationes
        Math., 89.1 (1986), 87-97.
[2]   Extensions of integral operators (to appear).

E. Lami Dozo and Ph. Turpin

[1]   Nonexpansive maps in generalized Orlicz spaces,
        Studia Math. 86.2 (1987), 155-188.

F. Leja

[1]   Sur les suites de polynomes bornée presque partout
        sur la frontière d'un domaine, Math.
        Ann. 108 (1933), 517-524.

G. Lewicki

[1]   On a theorem of S.N. Bernstein in metrizable
        topological linear spaces (to be published).

J. Lindenstrauss and J. Tzafriri

[1]   Classical Banach Spaces I, Sequence Spaces, Springer
        Verlag, Berlin--Heidelberg-New York 1977.
[2]   Classical Banach Spaces II, Function Spaces, Springer
        Verlag, Berlin--Heidelberg-New York 1979.

W.A.J. Luxemburg

[1]   Banach function spaces, Thesis, Delft 1955.

[2]   Notes on Banach function spaces XIV, Proc. Acad.
        Sci. Amsterdam (1965) A-68, 229-248.

[3]   Notes on Banach function spaces XV, Proc. Acad.
        Sci. Amsterdam (1965) A-68, 415-446.

[4]   Notes on Banach function spaces XVI, Proc. Acad.
        Sci. Amsterdam (1966) A-68, 664-667.

W.A.J Luxemburg and A.C. Zaanen

[1]-[12]   Notes on Banach function spaces I - XIII,
        Proc. Acad. Sci. Amsterdam,(1963) A-66, 135-153,
        239-263, 496-504, 655-681;(1964) A-64, 101-119;
        (1964) A-67, 360-376, 493-543.

[13]   Riesz spaces I, North Holland Amsterdam 1971.

J. Musielak

[1]   Orlicz Spaces and Modular Spaces, Lecture Notes in
        Mathematics 1034, Springer Verlag, Berlin-
        Heidelberg-New York 1983.

J. Musielak and W. Orlicz

[1]   On modular spaces, Studia Math. 18 (1959), 49-65.

[2]   Some remarks on modular spaces, Bull. Acad. Polon.
      Sci. Ser. Sci. Math. Astronom. Phys.7 (1959),
      661-668.

J. Musielak and A. Waszak

[1]   On a property of some methods of summability,
      Publ. Elektr. Fak. Univ. Beograd. 280
      (1969), 27-32.

[2]   Countably modulared spaces connected with
      equisplittable families of measures, Comment.
      Math. 13 (1970), 267-274.

[3]   On some countably modulared spaces, Studia Math.
      38 (1970), 51-57.

[4]   Some new countably modulared spaces, Comment.
      Math. 15 (1971), 209-215.

H. Nakano

[1]   Modulared semi-ordered linear spaces, Tokyo 1950.

W. Orlicz

[1]   Über eine gewisse klasse von Raumen vom Typus B,
      Bull. Acad. Polon. Sci. Ser. A (1932), 207-220.
[2]   Über Raumen L^M, Bull. Acad. Polon. Sci. Ser. A
      (1936), 93-107.

J. Peetre

[1]   A new approach in interpolation spaces, Studia
      Math. 34 (1970), 23-42.

W. Pleśniak

[1]   Quasianalytic functions in the sense of Bernstein,
      Dissertationes Mathematicae 147 (1977), 1-66.
[2]   Quasianalyticity in F-spaces of integrable functions,
      Approximation and Function Spaces,
      Ed. Z. Ciesielski, PWN Warszawa, North Holland,
      Amsterdam, New York, Oxford 1981.
[3]   Leja's type polynomial condition and polynomial
      approximation in Orlicz spaces, Ann.
      Polon. Math. 46 (1985), 268-278.

S. Reich

[1]   Extension problems for accretive sets in Banach
       spaces, J. Functional Analysis 26(1977), 378 - 395.

[2]   Personal communication, 1987.

S. Rolewicz

[1]   Metric linear spaces, 2nd edition,
       PWN Warszawa 1984.

R.L. Rosenberg

[1]   Orlicz spaces based on families of measures, Studia
       Math. 35 (1970), 15-49.

Ch. Shutao

[1]   Some rotundities of Orlicz spaces with Orlicz norm,
       Bull. Pol. Acad. Sci. Math. 43(9-10) (1986), 585-596.

Ch. Shutao et al

[1]   Geometry of Orlicz spaces, Harbin 1986.

J. Siciak

[1]   Extremal plurisubharmonic functions in C^n, Ann.
        Polon. Math. 39 (1981), 175-211.

[2]   On some extremal functions and their applications in
        the theory of analytic functions of several complex
        variables, Trans. Amer. Math. Soc. 105.2 (1962),
        322-357.

A. Szankowski

[1]   A Banach lattice without the approximation
        property, Israel J. Math. 24 (1976), 329-337.

T. Szczypiński

[1]   Nonlinear operator valued measures and integration,
        Comment. Math (in print).

P. Szeptycki

[1]   Notes on integral transformations, Dissertationes
       Math. 231 (1984), 1-52.

[2]   On some properties of domains of integral operators,
       Rocky Mountain J. of Math. 14.2 (1984), 433-440.

[3]   Personal communication, 1987.

B. Turett

[1]   Rotundity of Orlicz spaces, Indagationes Math., Ser.
       A, 79.5 (1976), 462-468.

[2]   Fenchel-Orlicz spaces, Dissertationes Math. 181
       (1980), 1-60.

A. Waszak

[1]   Some remarks on Orlicz spaces of strongly (A,ϕ)-
       summable sequences, Bull. Acad. Polon.Sci. Ser. Sci.
       Math. Astronom. Phys. 15 (1967), 265-269.

[2]   Strong summability of functions in Orlicz metrics,
       Comment. Math. 12 (1968), 115-139.

J.H. Wells and L.R. Williams

[1]   Embeddings and Extensions in Analysis, Springer-
        Verlag, Berlin, Heidelberg, New York, 1975

A.C. Zaanen

[1]   Linear analysis, Amsterdam-Groningen 1960.

Index